Series/Number 07-138

LOGIT AND PROBIT
Ordered and Multinomial Models

VANI K. BOROOAH
University of Ulster

SAGE PUBLICATIONS
International Educational and Professional Publisher
Thousand Oaks   London   New Delhi

For information:

Sage Publications, Inc.
2455 Teller Road
Thousand Oaks, California 91320
E-mail: order@sagepub.com

Sage Publications Ltd.
6 Bonhill Street
London EC2A 4PU
United Kingdom

Sage Publications India Pvt. Ltd.
M-32 Market
Greater Kailash I
New Delhi 110 048 India

Printed in the United States of America

Library of Congress Cataloging-in-Publication Data

Borooah, Vani K.
 Logit and probit: ordered and multinomial models / Vani K. Borooah.
  p. cm. – (Quantitative applications in the social sciences; v. 138)
  Includes bibliographical references.
  ISBN 0-7619-2242-3
  1. Social sciences–Statistical methods. 2. Probabilities. 3. Logits. 4. Probits.
  I. Title. II. Sage university papers series. Quantitative applications in the
social sciences; no. 07-138
  HA31.7 .B67   2001
  519.2–dc21   2001031926

This book is printed on acid-free paper.

01 02 03 04 05 06 10 9 8 7 6 5 4 3 2 1

Acquiring Editor:	C. Deborah Laughton
Editorial Assistant:	Veronica Novak
Production Editor:	Denise Santoyo
Production Assistant:	Kathryn Journey
Typesetter:	Technical Typesetting Inc.

When citing a university paper, please use the proper form. Remember to cite the Sage University Paper series title and include paper number. One of the following formats can be adapted (depending on the style manual used):

(1) BOROOAH, V. K. (2001) *Logit and Probit: Ordered and Multinomial Models.* Sage University Papers Series on Quantitative Applications in the Social Sciences, 07-138. Thousand Oaks, CA: Sage.

OR

(2) Borooah, V. K. (2001). *Logit and Probit: Ordered and Multinomial Models.* (Sage University Papers Series on Quantitative Applications in the Social Sciences, series no. 07-138). Thousand Oaks, CA: Sage.

CONTENTS

SERIES EDITOR'S INTRODUCTION

For ordinary least squares (OLS) to yield BLUE estimators, the classical regression assumptions must be met. Some of these assumptions are easier to meet than others. Further, the substantive consequences of their violation vary, from assumption to assumption. One assumption that can be hard to meet, and that has serious consequences for OLS interpretation if not met, is the assumption that the dependent variable is continuous. If instead the dependent variable is discrete, consisting of two or more outcome categories, then OLS poses serious inference problems. In such circumstances, maximum likelihood techniques such as logit or probit are generally more efficient. The Sage QASS series has given considerable attention to application of logit or probit when the dependent variable is dichotomous. Consult the relevant sections in No. 45, *Linear Probability, Logit, and Probit Models*, by Aldrich and Nelson; No. 86, *Logit Modeling*, by DeMaris: No. 101, *Interpreting Probability Models: Logit, Probit, and Other Generalized Linear Models*, by Liao; No. 106, *Applied Logistic Regression Analysis*, by Menard; and No. 132, *Logistic Regression: A Primer*, by Pampel. While these works may touch on discrete dependent variables with categories > 2, they do not emphasize them.

The monograph at hand is unique, because it attends exclusively to estimation when the dependent variable has multiple categories. After an introduction, dependent variables that are discrete and ordered are considered. For example, suppose a political scientist has election survey data, and wishes to explain the dependent variable of Political Interest, where respondents are scored 0=low, 1=moderate, 2=high. The variable is discrete, with respondents falling into one of three categories. Further, the variable is ordered from "low" to "high" interest. With such an ordinal variable, we might say that someone who scored "high" has more political interest than someone scored "low," but we cannot say precisely how much more. Thus, OLS regression seems less desirable, ordered logit or ordered probit more desirable, for they accommodate this lower measurement level. Professor Borooah explicates both procedures in an effort to account for individual differences in social deprivation (measured in three categories, "not deprived," "mildly deprived," and "severely deprived"). One question that often comes up is whether logit is preferred over probit, or vice-versa. The fundamental theoretical difference between the two approaches concerns the distribution of the error term, logistic versus normal. In practice, as noted here, it is difficult to justify the selection of one over the other.

Treatment is eventually extended to multinomial, or non-ordered, dependent variables with categories > 2. For example, choice of religion, choice of neighborhood, choice of shopping center, choice of job. A key assumption of multinomial logit is the Independence of Irrelevant Alternatives (IIA). As Professor Borooah discusses, this assumption is both the strength and the weakness of the technique. He also makes the important but often forgotten distinction between odds-ratios and risk-ratios. With binary logit there is no difference between the two ratios; however, with multinomial logit outcomes are given in terms of risk-ratios.

The text concludes with helpful detail on the computer programs actually used to obtain the table results. This step-by-step explication of computing procedure allows readers to see how to run the analyses. The exposition is in STATA, but the author also points out other available software in SAS, SPSS, and LIMDEP. Overall, the monograph provides a current guide to estimating and interpreting results from the more complex discrete dependent variable models.

—*Michael S. Lewis-Beck*
Series Editor

PREFACE

There are many instances where the appropriate variable for analysis is merely a coding for some qualitative outcome. As a consequence, in such situations the dependent variable takes a discrete number of mutually exclusive, and collectively exhaustive, values. This is in contrast to other situations in which, at least conceptually, the dependent variable assumes a continuum of values. Although conventional regression methods are not appropriate for the statistical analysis of discrete dependent variables, one can, nonetheless, in the spirit of regression analysis, construct models which link the observed outcome to the values of certain "determining" or "explanatory" variables. This monograph discusses the estimation, simulation and interpretation of multiple (> 2) outcome models, with ordered and with unordered outcomes, against the backdrop of questions relating to socioeconomic inequality.

In preparing this monograph, I am grateful to Michael Lewis-Beck and to two anonymous referees for their valuable comments. I am also grateful to the Social Policy Association for permission to reproduce material from my paper, "Targeting social need: Why are deprivation levels in Northern Ireland higher for Catholics than for Protestants?" (Borooah, 2000) and to the Scottish Economic Society for permission to reproduce material from my paper, "How do employees of ethnic origin fare on the occupational ladder in Britain?" (Borooah, 2001). The results reported in this monograph are based on data from the 1991 Census for Great Britain. This data, which is Crown copyright, was kindly made available by the Census Microdata Unit at the Cathy Marsh Centre for Census and Survey Research, University of Manchester, through funding by JISC/ESRC/DENI. Needless to say, however, I alone remain responsible for the results, their interpretation and, indeed, for any errors that this work might contain.

LOGIT AND PROBIT: ORDERED AND MULTINOMIAL MODELS

VANI K. BOROOAH
University of Ulster

1. INTRODUCTION

Kenneth Tynan, the theatre critic, once famously described his profession as consisting of people "who know the way, but can't drive the car." Researchers in the social sciences often find themselves in an analogous situation. Some can locate points on their research map but are uncertain how to get from one place to another. Others can trace their path through the suburbia of research but feel less confident of getting behind the steering wheel. A fortunate few can both navigate and pilot. This observation defines the broad purposes of this monograph, which are (a) to trace the main paths that lead through the landscape of *ordered* and *multinomial* logit models, and (b) to offer instruction in driving along these routes. But first, before gears are engaged, some words by way of introduction.

There are many instances where the appropriate variable for analysis is merely a coding for some qualitative outcome. Such models are known as *qualitative choice* models. For example, in judging a government's performance, a person strongly approves (coding = 1); approves (coding = 2); disapproves (coding = 3); or strongly disapproves (coding = 4). Or, a person votes Liberal (coding = 1), Conservative (coding = 2), or Labour (coding = 3). As a consequence, in such situations, the dependent variable takes a *discrete* number of mutually exclusive, and collectively exhaustive, values. This is in contrast to other situations in which, at least conceptually, the dependent variable assumes a continuum of values. Although conventional regression methods are not appropriate for the statistical analysis of discrete dependent variables, one can, nonetheless, in the spirit of regression analysis, construct models which link the observed outcome to the values of certain "determining" or "explanatory" variables. Qualitative choice models in which the dependent variable takes more

1

than two values are known as *multiple outcome* models. One may further subdivide the class of multiple outcome models into those involving (a) ordered outcomes (such as the example on the degree of approval of governmental performance, above) and (b) unordered outcomes (such as the voting example, above).

Qualitative choice models have become a growth industry in applied econometric analysis. Social scientists have always had an interest in the choice between mutually exclusive options. The increasing availability of survey data (in either cross-section or panel form) has meant that more and more scholars are able to translate their intellectual speculation into hard results. Scholars come to these data with a specific set of questions to which they seek answers. However, when these data are analyzed, the results are often shrouded in a Delphic obliqueness about what the "right" answers might be. In order to dispel this fog of ambiguity, textbook techniques have to be manipulated so that the derived results point clearly in the right direction. This monograph discusses the *estimation, simulation* and *interpretation* of multiple outcome models, with ordered and unordered outcomes, against the backdrop of questions relating to socioeconomic inequality.

Ordered and unordered models require different techniques for their respective analysis. Ordered models may be estimated by either *logit* methods, which are known as *ordered logit* models, or by *probit* methods, which are known as *ordered probit* models. Models where the outcomes are unordered are most easily estimated by logit methods. Although in principle it is possible to estimate such models by probit, for computational reasons it is often not feasible. For that reason, multiple outcome models with unordered outcomes are referred to as *multinomial logit* models.

Multinomial logit models may be *conditional*, which means that the choices between alternatives may depend not just upon the *characteristics* of the individual making the choice but also upon the *attributes* of the choice. For example, the choice by individuals of which shopping center to patronize may depend upon the attributes of the centers (number and variety of shops, the standard of upkeep of the centers), which do not vary across the individuals, and upon income and family size, which do vary across the individuals. A further complication is that individual characteristics and choice attributes may *interact*: the choice of shopping center may depend on the distance which an individual has to travel from his/her place of residence (residence being an individual's characteristic) to a particular shopping center (the location of the center being a choice attribute).

Ordered models are discussed in Chapter 2 and multinomial logit models are discussed in Chapter 3. Both chapters are united by two aims, (a) to convey an understanding of the underlying methodology of the models and (b) to impart an ability to use the models for research in the social sciences.

Meeting the first aim involves steering a safe path between the Scylla of oversimplification and the Charybdis of excessive technicality. I have tried to navigate this narrow passage without, I hope, holing the ship. However, in order to lighten its load, I have assumed that the reader has sufficient knowledge of the material that forms the "prequel" to this monograph. In particular, these include:

(1) The deficiencies of the linear probability model, or why the methods of ordinary regression analysis are not appropriate for analyzing models with discrete dependent variables.

(2) A familiarity with logit and probit methods as applied to models where the dependent variable has only *two* possible outcomes.

For those wishing to refresh their memories on these topics, earlier monographs in this series by Aldrich and Nelson (1984), DeMaris (1992), Liao (1994) and Menard (1995), as well as more general econometrics texts like Greene (2000), provide an excellent review.

Meeting the second aim is, in my view, more difficult. Using a model involves several layers of understanding. First, one needs to be clear about the questions to ask. Next, there is the problem of how to answer the questions. Typically, a particular research question can be answered in more than one way, and so it is important to know how these different ways differ, why they differ, and what might be the best way for addressing the problem at hand. Lastly, after one has decided on the questions and how they are to be answered, there is the practical problem of going about obtaining the answers—of implementing one's research strategy by obtaining and interpreting results.

I have tried to address these issues—in the context of ordered logit and probit, and multinomial logit, models—by adopting a three-fold strategy. First, I have tried to secure the maximum possible overlap between the exposition of the methodology and the empirical analysis. There is very little that is set out in the theoretical sections of any chapter that is not echoed in the sections of that chapter concerned with the application of the methodology.

Second, I have anchored the exposition in two major pieces of applied work. I did so because I felt that having a single empirical (and "real world") thread running through a chapter might better illuminate the use of a model than the separate strands of fragmented examples. Chapter 2, on ordered logit and probit, is grounded in empirical work on deprivation. This work uses data from the Northern Ireland census on nearly 14,000 individuals to examine where the roots of deprivation and inequalities in the deprivation experience between Catholics and Protestants might lie. Chapter 3, on multinomial logit models, is rooted in empirical work on occupational attainment by ethnic minorities in Britain. This work uses data from the British census on nearly 100,000 male full-time employees to examine the different chances of black Caribbeans, Indians, and whites of being in various occupational categories.

Third, the last chapter contains a complete listing of the computer programs used to generate the empirical results. There are, to my knowledge, at least four well-known and highly regarded pieces of software which, among other things, handle problems of the kind discussed in this monograph: *SAS; SPSS v10.0* (a good introduction to the use of SAS and SPSS procedures in the analysis of events with ordered and unordered outcomes is provided by the Stat/Math Center at the Indiana University[1]); *LIMDEP* (see Greene, 1995); and *STATA* (see STATA, 1999). It just so happens that, by a quirk of fate, I am most familiar with STATA. For this reason the programs of Chapter 4 are written in STATA code. Almost every line in these programs has a comment attached to it that explains what it is supposed to do and how it relates to the material of the earlier chapters. The earlier chapters map the route; Chapter 4 teaches how to drive the car!

2. ORDERED MODELS

Introduction

Suppose that there are N persons (indexed $i = 1, \ldots, N$) for each of whom an "event" can occur. Suppose that this event has $M > 2$ outcomes, indexed $j = 1, \ldots, M$, where these outcomes are mutually exclusive and collectively exhaustive. Let the values taken by the variable Y_i represent these outcomes for person i such that: $Y_i = 1$ if the first outcome occurs for this person ($j = 1$); $Y_i = 2$ if the second outcome occurs ($j = 2$) and so on till $Y_i = M$ if the last outcome occurs ($j = M$). Suppose further that these outcomes are *inherently ordered*, by which is meant that the outcome associated with a higher value of the variable Y_i *is ranked higher* than the outcome associated with a lower value of the variable. Another way of expressing this is to say that the dependent variable, Y_i, associated with the outcomes is *ordinal*: "stronger" outcomes are associated with higher values of the variable. However, this ordinal nature of the outcomes has no implication for differences in the strength of the outcomes; the outcome associated with $Y_i = 2$ is not twice as strong as that associated with $Y_i = 1$. Consequently, the actual values taken by an ordinal dependent variable are irrelevant, so long as larger values correspond to stronger outcomes: we could have defined $Y_i = 5$ if the first outcome occurred; $Y_i = 7$ if the second outcome occurred and so on.

An example of ordered outcomes is provided by a person's health status. The outcomes associated with this, e.g., "poor," "good," and "excellent," could be represented by a variable taking, respectively, the values 1, 2 and 3. In this example, the outcome associated with $Y_i = 3$ (excellent health) is better than that associated with $Y_i = 2$ (good health) and this, in turn, is better than the outcome associated with $Y_i = 1$ (poor health). Other examples of ordered outcomes are outcomes relating to the level of insurance coverage taken by a person (no cover, part cover, full cover) or outcomes relating to the employment status of working-age persons (inactive, unemployed, employed). One example of a *nonordered* outcome is a person's religion, e.g., $Y_i = 1$ for Christians, $Y_i = 2$ for Jews, $Y_i = 3$ for Muslims, and $Y_i = 4$ for Hindus. While the outcomes are all different in this example, they cannot be ranked and, therefore, cannot be regarded as ordered outcomes. To put it differently, the dependent variable associated with the religion outcomes is not ordinal.

When the outcomes are clearly ordered, one should take account of the fact that the dependent variable is both discrete and ordinal. For example, if the outcomes are coded 1, 2, 3, a linear regression would treat the difference between a 3 and a 2 identically to the difference between a 2 and a 1 whereas, in fact, the numbers are only a ranking and have no cardinal significance. On the other hand, to estimate an econometric relation with an ordinal dependent variable using the methods of multinomial logit (discussed in the next chapter) would mean that the information conveyed by the ordered nature of the data was being discarded. The most commonly used and appropriate methods for estimating models *with more than two outcomes*,[2] when the dependent variable associated with the outcomes is both discrete and ordinal, are those of *ordered logit* and *ordered probit*.[3]

However, the above observation is subject to an important caveat. A critical assumption of ordered logit and probit is that of *parallel slopes*. The implications of this assumption are discussed in some detail below, but in essence it means that if there is a variable which affects the likelihood of a person being in the ordered categories (e.g., diet on health status) then it is assumed that the coefficients linking the variable value to the different outcomes will be the same across all the outcomes (a given diet will affect the likelihood of a person being in excellent health exactly as it will affect the likelihood of him or her being in poor health). If this assumption is invalid, so that the slope coefficients associated with a particular variable are different across the different outcomes (a given diet will affect the likelihood of a person being in excellent health differently than it will affect the likelihood of him or her being in poor health), then the methods of ordered logit and probit are no longer appropriate and the model ought to be estimated using the methods of multinomial logit (discussed in Chapter 3).

The fact that it is not always possible to unambiguously identify outcomes as ordinal provides another reason for being cautious in the use of ordered estimation methods. For example, where a person lives in a city (North, South, East, or West) is ostensibly a nonordered variable. But if one knew that certain parts of the city provided more salubrious living conditions than other parts, then a variable defining a person's place of residence could acquire an ordinal connotation with, say, living in the North ($Y_i = 4$) being "better" than living in the South ($Y_i = 3$). However, in the face of uncertainty about whether a variable is ordered or nonordered, a sensible rule might be to regard it as nonordered and, as a corollary, to estimate models using it as a dependent variable by the methods of multinomial logit. This rule is sensible because treating an outcome variable as ordered, when in fact it is nonordered, imposes a ranking on the outcomes that they do not possess and invokes the restrictive assumption of parallel slopes (referred to above and discussed in detail below), which is likely to bias the estimates. On the other hand, not treating an outcome variable as ordered, when in fact it is ordered, fails to impose a legitimate ranking on the outcomes. This omission may lead to a loss of efficiency, but it is unlikely to bias the estimates. In the face of these two possible errors, the loss of efficiency a less serious error to make than that of biased estimates. Another example of such ambiguity—and one that drives the next chapter on multinomial logit—is provided by a person's occupation. Whether a person works in an unskilled/semi-skilled, skilled, or professional/managerial job could be regarded as a matter of individual choice, although the constraints that affect this choice may differ from person to person and vary according to race and/or gender inter alia. On this choice of interpretation, occupational outcomes could be viewed as "nonordered," meaning there is nothing inherently desirable or undesirable in one type of occupational category over another. For this interpretation, the appropriate estimation method for analyzing occupational outcomes is multinomial logit. Indeed, this was precisely the method adopted by Schmidt and Strauss (1975) in their analysis of the occupation of 1,000 persons in terms their education, experience, race and sex. This method was echoed in Greene (2000, p. 859) in providing an example of the use of multinomial logit.

On the other hand, if a university professor was asked whether he or she would prefer to have a banker or a janitor as his son-in-law, that professor might plump for the former. In expressing this preference, the professor is implicitly "ranking" occupations, with bankers being in higher ranked jobs (making more desirable sons-in-law) than janitors. However, the point is that this ranking is purely subjective[4] (i.e., there is nothing inherently more desirable about being a banker than about being a janitor) and does not carry the objectivity that would be attached to the professor's preference that his or her son-in-law should enjoy good health rather than suffer ill health. The moral of the story is that it is better to treat outcomes as nonordered *unless one has good reasons for imposing a ranking.*[5]

Methodology

The methodology and underlying logic of ordered logit and probit models are perhaps best presented using a concrete example. Suppose that there are N persons (indexed $i = 1, \ldots, N$) living in an area and that each person's "degree of deprivation" can be represented by the value of a variable D_i, such that higher values of D_i represent higher degrees of deprivation. The value assumed by this "deprivation index" for a particular person—hereafter referred to as his or her "deprivation score"—depends upon a variety of factors pertaining to that person. Examples of such factors might include being unemployed, being a single parent, and living in a particular area. Suppose that the deprivation index, D_i, is a *linear* function of K factors ("determining variables") whose values, for individual i, are X_{ik}, $k = 1, \ldots, K$. This means that the deprivation index can be represented as

$$D_i = \sum_{k=1}^{K} \beta_k X_{ik} + \varepsilon_i = Z_i + \varepsilon_i. \tag{2.1}$$

where β_k is the coefficient associated with the kth variable ($k = 1, \ldots, K$) and $Z_i = \sum_{k=1}^{K} \beta_k X_{ik}$. An increase in the value of the kth factor for a particular person will cause his or her deprivation score to rise if $\beta_k > 0$ and fall if $\beta_k < 0$. However, because the relationship between the deprivation score and the deprivation-inducing factors is not an exact one—for example, there may be factors left out of the equation or factors may be measured inaccurately—an error term, ε_i is included in the equation to capture this inexactitude.

The problem with the formulation in Equation 2.1 is that the exact shade of a person's deprivation, as represented by the values of D_i, is difficult, if not impossible, to observe. The deprivation index is a *latent variable*, which (though conceptually useful) is unobservable either in principle or practice, and Equation 2.1 is a *latent regression*, which as it stands cannot be estimated.

However, what can be observed is a person's deprivation *level*—a person can, for example, be classified as being "not deprived," "mildly deprived," or "severely deprived"—and a variable Y_i can be associated with these deprivation levels, such that $Y_i = 1$ if person is not deprived, $Y_i = 2$ if person is mildly deprived, and $Y_i = 3$ if person is severely deprived. In terms of the earlier discussion, Y_i is an ordinal variable. The categorization of the persons in the sample in terms of the three levels of deprivation is *implicitly* based on the values of the latent variable D_i, in conjunction with "threshold" values δ_1 and δ_2, such that

$$
\begin{aligned}
Y_i &= 1, \quad \text{if } D_i \le \delta_1 \\
Y_i &= 2, \quad \text{if } \delta_1 \le D_i \le \delta_2 \\
Y_i &= 3, \quad \text{if } D_i \ge \delta_2.
\end{aligned}
\tag{2.2}
$$

The $\delta_1, \delta_2 \geq 0$ of Equation 2.2 are unknown parameters ($\delta_1 < \delta_2$) to be estimated along with the β_k of Equation 2.1. A person's classification in terms of deprivation level depends upon whether or not his or her deprivation score, D_i, crosses a threshold. The probabilities of Y_i taking values 1, 2 and 3 are given by

$$\Pr(Y_i = 1) = \Pr(Z_i + \varepsilon_i \leq \delta_1) = \Pr(\varepsilon_i \leq \delta_1 - Z_i)$$
$$\Pr(Y_i = 2) = \Pr(\delta_1 \leq Z_i + \varepsilon_i \leq \delta_2) = \Pr(\delta_1 - Z_i < \varepsilon_i \leq \delta_2 - Z_i) \quad (2.3)$$
$$\Pr(Y_i = 3) = \Pr(Z_i + \varepsilon_i \geq \delta_2) = \Pr(\varepsilon_i \geq \delta_2 - Z_i).$$

Each of the N observations is treated as a single draw from a multinomial distribution, and in this case the multinomial distribution has three outcomes, not deprived, mildly deprived, and severely deprived. Suppose that of the N persons, N_1 were not deprived, N_2 were mildly deprived and N_3 were severely deprived.[6] Then the likelihood of observing the sample, which is simply the product of the probability of the individual observations, is

$$L = [\Pr(Y_i = 1)]^{N_1} [\Pr(Y_i = 2)]^{N_2} [\Pr(Y_i = 3)]^{N_3}$$
$$= [F(\delta_1 - Z_i)]^{N_1} [F(\delta_2 - Z_i) - F(\delta_1 - Z_i)]^{N_2}$$
$$\times [1 - F(\delta_1 - Z_i)]^{N_3}. \quad (2.4)$$

where $F(x) = \Pr(\varepsilon_i < x)$ is the cumulative probability distribution of the error terms. If we *knew* the probability distribution of the error terms—that is, if we knew what $F(x)$ was—then we could choose as our estimates of β_k, δ_1, and δ_2 those values which *maximized the likelihood* of observing the sample observations.[7] In the absence of such knowledge, we could *assume* that the error terms followed a particular probability distribution.

The difference between the *ordered logit* and the *ordered probit* models lies in the (assumed) distribution of ε_i, the error term in Equation 2.1. An ordered logit model is the result of assuming that ε_i is *logistically* distributed, while an ordered probit model is the result of assuming that ε_i is *normally* distributed. It is natural to ask which distribution is the appropriate one to use.[8] The logistic distribution is similar to the normal except in the tails, which are considerably heavier.[9] As Greene (2000) points out, "it is difficult to justify the

choice of one distribution over the other on theoretical grounds...
in most applications, it seems not to make much difference" (p. 815).

Using the estimated values $\hat{\beta}_k$ of the coefficients β_k allows an estimated value $\hat{Z}_i = \sum_{k=1}^{K} \hat{\beta}_k X_{ik}$ to be computed for each individual in
the sample. Using the $\hat{Z}_i$ in conjunction with $\hat{\delta}_1$ and $\hat{\delta}_2$, which are the
estimated values of the cutoff parameters δ_1 and δ_2, allows the probabilities of being at different levels of deprivation to be estimated for
every person in the sample. These estimates—respectively denoted
$\hat{p}_{i1}$, $\hat{p}_{i2}$, and $\hat{p}_{i3}$—are computed as

$$\hat{p}_{i1} = \Pr(\varepsilon_i \leq \hat{\delta}_1 - \hat{Z}_i) = F(\hat{\delta}_1 - \hat{Z}_i) \tag{2.5a}$$

$$\hat{p}_{i2} = \Pr(\hat{\delta}_1 - \hat{Z}_i < \varepsilon_i \leq \hat{\delta}_2 - \hat{Z}_i) = F(\hat{\delta}_2 - \hat{Z}_i) - F(\hat{\delta}_1 - \hat{Z}_i) \tag{2.5b}$$

$$\hat{p}_{i3} = \Pr(\varepsilon_i \geq \hat{\delta}_2 - \hat{Z}_i) = 1 - F(\hat{\delta}_2 - \hat{Z}_i) \tag{2.5c}$$

where $\sum_{j=1}^{3} \hat{p}_{ij} = 1$ for all $i = 1, \ldots, N$.

The model described above is also known as the *proportional-odds*
model, because if one considers the odds-ratio with respect to some
category $j = m$,

$$OR(m) = \frac{\Pr(Y_i \leq m)}{\Pr(Y_i > m)},$$

then this ratio is independent of the category m. The odds ratio is
assumed to be constant for all categories.[10]

A Clarification on Notation

The ordered regressions as estimated by STATA (whether logit or
probit) *do not explicitly include* an intercept term: in other words,
the β_k $(k = 1, \ldots, K)$ in Equation 2.1 are all slope coefficients.
The intercept term is not explicitly shown because it is absorbed,
in a manner to be shown below, into the cutoff points, δ_1 and δ_2.
On the other hand, Greene (2000, p. 876), in his formulation of the
ordered regression, *explicitly includes* an intercept term. His equivalent of Equation 2.1 is

$$D_i = \beta_0 + \sum_{k=1}^{K} \beta_k X_{ik} + \varepsilon_i = \beta_0 + Z_i + \varepsilon_i = W_i + \varepsilon_i \tag{2.6}$$

where β_0 is the intercept term and $W_i = \beta_0 + Z_i$. Greene's cutoff points are denoted by μ_1 and μ_2 (which are different from STATA's cutoff points δ_1 and δ_2; precisely how they are different is shown below). Greene (2000) then sets the first cutoff point, μ_1, to zero.[11] Therefore, under his formulation, the equations for the probabilities $\hat{p}_{i1}$, $\hat{p}_{i2}$, and $\hat{p}_{i3}$ become

$$\hat{p}_{i1} = \Pr(\varepsilon_i \leq -\widehat{W}_i) = F(-\widehat{W}_i) = F(-\hat{\beta}_0 - \widehat{Z}_i)$$
$$= F(\hat{\delta}_1 - \widehat{Z}_i) \tag{2.7a}$$

$$\hat{p}_{i2} = \Pr(-\widehat{W}_i < \varepsilon_i \leq \hat{\mu}_2 - \widehat{W}_i) = F(\hat{\mu}_2 - \widehat{W}_i) - F(-\widehat{W}_i)$$
$$= F(\hat{\mu}_2 - \hat{\beta}_0 - \widehat{Z}_i) - F(-\hat{\beta}_0 - \widehat{Z}_i)$$
$$= F(\hat{\delta}_2 - \widehat{Z}_i) - F(\hat{\delta}_1 - \widehat{Z}_i) \tag{2.7b}$$

$$\hat{p}_{i3} = \Pr(\varepsilon_i \geq \hat{\mu}_2 - \widehat{W}_i) = 1 - F(\hat{\mu}_2 - \widehat{W}_i)$$
$$= 1 - F(\hat{\mu}_2 - \hat{\beta}_0 - \widehat{Z}_i) = 1 - F(\hat{\delta}_2 - \widehat{Z}_i). \tag{2.7c}$$

Equations 2.5a to 2.5c of STATA and Equations 2.7a to 2.7c of Greene (2000) are equivalent[12] when $\delta_i = \mu_i - \beta_0$. That is to say, STATA cutoff points are equal to Greene's cutoff points *less* the intercept term. In that sense, STATA absorbs the intercept term into its cutoff points.[13]

Ordered Logit

Under a logistic distribution, the cumulative distribution function of the random variable X is

$$\Pr(X \leq x) = \Lambda(x) = \exp(x)/[1 + \exp(x)] = 1/(1 + \exp(-x)), \tag{2.8}$$

and so if it is assumed that the error terms follow a logistic distribution,

$$\Pr(Y_i = 1) = \Lambda(\delta_1 - Z_i) = 1/[1 + \exp(Z_i - \delta_1)] \tag{2.9a}$$

$$\Pr(Y_i = 2) = \Lambda(\delta_2 - Z_i) - \Lambda(\delta_1 - Z_i)$$
$$= 1/[1 + \exp(Z_i - \delta_2)] - 1/[1 + \exp(Z_i - \delta_1)] \tag{2.9b}$$

$$\Pr(Y_i = 3) = 1 - \Lambda(\delta_2 - Z_i) = 1 - 1/[1 + \exp(Z_i - \delta_2)]. \tag{2.9c}$$

The estimates of the β_k, δ_1 and δ_2 are obtained by maximizing the likelihood function (Equation 2.4), using the logistic distribution function $\Lambda(.)$ in place of $F(.)$.

Ordered Probit

The cumulative distribution of a standard normal variate[14] (SNV) X is

$$\Pr(X < x) = \Phi(x) = \int_0^x (1/2\pi)\exp(-X^2/2)dX, \qquad (2.10)$$

and so if it is assumed that the error terms are SNVs

$$\Pr(Y_i = 1) = \Phi(\delta_1 - Z_i) \qquad (2.11a)$$

$$\Pr(Y_i = 2) = \Phi(\delta_2 - Z_i) - \Phi(\delta_1 - Z_i) \qquad (2.11b)$$

$$\Pr(Y_i = 3) = 1 - \Phi(\delta_2 - Z_i). \qquad (2.11c)$$

The estimates of the β_k, δ_1 and δ_2 are obtained by maximizing the likelihood function (Equation 2.4), using the normal distribution function $\Phi(.)$ in place of $F(.)$.

Marginal Effects: Continuous Variables

A natural question to ask is how the probabilities of the various outcomes would change when the value of one of the variables influencing the outcomes changes. For example, if age is a factor which influences deprivation then how would a person's probability of being at the different deprivation levels (not deprived, mildly deprived, and severely deprived) be affected if he or she was a year older or younger? The marginal effect on the three probabilities for person i, of a small change in X_{ik} (the value of the kth determining variable for person i), under a logistic distribution, is

$$\frac{\partial \Pr(Y_i = 1)}{\partial X_{ik}} = \frac{d}{dZ_i}[\Lambda(\delta_1 - Z_i)]\frac{\partial Z_i}{\partial X_{ik}} = -\Lambda'(\delta_1 - Z_i)\beta_k \qquad (2.12a)$$

$$\frac{\partial \Pr(Y_i = 2)}{\partial X_{ik}} = \frac{d}{dZ_i}[\Lambda(\delta_2 - Z_i) - \Lambda(\delta_1 - Z_i)]\frac{\partial Z_i}{\partial X_{ik}}$$

$$= [\Lambda'(\delta_2 - Z_i) - \Lambda'(\delta_1 - Z_i)]\beta_k \qquad (2.12b)$$

$$\frac{\partial \Pr(Y_i = 3)}{\partial X_{ik}} = \frac{d}{dZ_i}[1 - \Lambda(\delta_2 - Z_i)]\frac{\partial Z_i}{\partial X_{ik}}$$

$$= \Lambda'(\delta_2 - Z_i)\beta_k, \qquad (2.12c)$$

and under a normal distribution is

$$\frac{\partial \Pr(Y_i = 1)}{\partial X_{ik}} = \frac{d}{dZ_i}[\Phi(\delta_1 - Z_i)]\frac{\partial Z_i}{\partial X_{ik}} = -\Phi'(\delta_1 - Z_i)\beta_k \qquad (2.13a)$$

$$\frac{\partial \Pr(Y_i = 2)}{\partial X_{ik}} = \frac{d}{dZ_i}[\Phi(\delta_2 - Z_i) - \Phi(\delta_1 - Z_i)]\frac{\partial Z_i}{\partial X_{ik}}$$

$$= [\Phi'(\delta_2 - Z_i) - \Phi'(\delta_1 - Z_i)]\beta_k \qquad (2.13b)$$

$$\frac{\partial \Pr(Y_i = 3)}{\partial X_{ik}} = \frac{d}{dZ_i}[1 - \Phi(\delta_2 - Z_i)]\frac{\partial Z_i}{\partial X_{ik}} = \Phi'(\delta_2 - Z_i)\beta_k \qquad (2.13c)$$

where $\Lambda'(x) = d\Lambda(x)/dx$ and $\Phi'(x) = d\Phi(x)/dx$ are the probability density functions of the logistic and of the normal distributions, respectively. The marginal effects can be obtained by evaluating the appropriate density functions at the relevant points and multiplying by the associated coefficient. For example, from Equation 2.8, the density function of the logistic distribution is

$$\Lambda'(x) = \frac{d}{dx}\left[\frac{\exp(x)}{1 + \exp(x)}\right] = \frac{[1 + \exp(x)]\exp(x) - [\exp(x)]^2}{[1 + \exp(x)]^2}$$

$$= \Lambda(x)[1 - \Lambda(x)].$$

The marginal effect under Equation 2.12a is given by

$$\Lambda'(\delta_1 - Z_i)\beta_k = \Lambda(\delta_1 - Z_i)[1 - \Lambda(\delta_1 - Z_i)]\beta_k$$

$$= \frac{1}{1 + \exp(Z_i - \delta_1)}\left[1 - \frac{1}{1 + \exp(Z_i - \delta_1)}\right]\beta_k.$$

Now if the value of the kth determining variable increases by a small amount and $\beta_k > 0$, then under both the logit and the probit model the probability of not being deprived must fall because, by Equations 2.12a and 2.13a, the derivative of $\Pr(Y_i = 1)$ has the *opposite* sign to β_k. Under both models too, the probability of being severely deprived must rise since by Equations 2.12c and 2.13c the derivative of $\Pr(Y_i = 3)$ has the *same* sign as β_k. However, it is not clear what would happen to the middle probability. Depending upon how the other probabilities change, the probability of being mildly deprived, $\Pr(Y_i = 2)$, could either rise, fall, or remain unchanged.[15] Consequently, given a change in a determining variable, it is impossible to infer the direction of change in *all* the probabilities from the sign of the coefficient associated with it. It is only the direction of change in the probabilities of the two extreme cases that will be unambiguously determined. For this reason, Greene (2000) cautions that "we must be very careful in interpreting the coefficients in this model... since this is the least obvious of the models" (p. 878).

Marginal Effects: Dummy Variables

However, the above approach for evaluating marginal effects is only appropriate when the determining variable is continuous and not when it is a dummy variable. The effects of a dummy variable should be analyzed by comparing the probabilities that result when the dummy variable takes one value with the probabilities that are the consequence of it taking the other value, the values of the other variables remaining unchanged between the two comparisons. For example, suppose employment status is a factor which influences deprivation and let $X_{ik} = 1$ if a person is unemployed and $X_{ik} = 0$ if he or she is employed. In order to analyze how a person's probabilities of being at the three deprivation levels would be affected if he or she moved from employment into unemployment, first evaluate Z_i under the assumption that $X_{ik} = 1$ (call it Z_i^1) and use Equations 2.9a to 2.9c—or, if a probit model is used, Equations 2.11a to 2.11c—to calculate the three probabilities for $Y_i = 1$ (not deprived), $Y_i = 2$ (mildly deprived), and $Y_i = 3$ (severely deprived). Then, keeping the values of the other determining variables unchanged, evaluate Z_i under the assumption that $X_{ik} = 0$ (call it Z_i^0) and recompute the three probabilities. Note that from Equation 2.1, $Z_i^1 = Z_i^0 + \beta_k$. The difference between the two sets of probabilities is the effect of a person moving from employment ($X_{ik} = 0$) into unemployment ($X_{ik} = 1$), or vice-versa, on his or her probability of being at different deprivation levels.

The Parallel Slopes Assumption

A critical assumption of the ordered logit and probit models is that the slope coefficients β_k of Equation 2.1 *do not vary* according to the deprivation outcome being considered. That is to say, the ordered logit and probit models fit a *parallel slopes cumulative model* which, in the logit case, takes the following form:

$$\log \frac{p_1}{1 - p_1} = \alpha_1 + \sum_{k=1}^{K} \beta_k X_{ik}$$

$$\log \frac{p_1 + p_2}{1 - p_1 - p_2} = \alpha_2 + \sum_{k=1}^{K} \beta_k X_{ik}$$

$$\cdots \cdots$$

$$\log \frac{p_1 + p_2 + \cdots + p_{M-1}}{1 - p_1 - p_2 - \cdots - p_{M-1}} = \alpha_M + \sum_{k=1}^{K} \beta_k X_{ik},$$

$$p_1 + p_2 + p_M = 1$$

where $p_j = \Pr(Y_i = j)$. The validity of the *parallel slope assumption* can be tested by estimating a multinomial logit model on the data (see chapter 3). The multinomial logit model allows the slope coefficients β_k *to be different* between the outcomes $j = 1, \ldots, M$. While the ordered logit model estimates K coefficients, the multinomial logit model estimates $K(M - 1)$ parameters.[16] If L_1 is the likelihood value from the ordered logit model and L_2 is the likelihood value from the multinomial logit model, then one can compute $2(L_2 - L_1)$ and compare with $\chi^2(K(M - 2))$. Note that this is not strictly a likelihood-ratio test because the ordered logit model is not nested within the multinomial logit model. Consequently, the test is only *suggestive*: a "very large" χ^2 value would provide grounds for concern, a "moderately large" value would not (STATA, 1999, p. 480). However, if one does have reason for believing that the parallel slope assumption is not valid then the model ought to be estimated using the method of multinomial logit, notwithstanding the fact that the dependent variable is clearly ordinal.

Application to Deprivation Status

Defining and Constructing the Deprivation Index

There are N persons, indexed, $i = 1, \ldots, N$. A condition is defined as a "deprivation-inducing condition" (DIC) if the presence of that condition causes an individual to experience deprivation. Suppose there are K DICs, indexed $k = 1, \ldots, K$ and let I_{ik} be a *categorical* variable with respect to DIC k and person i, such that $I_{ik} = 1$ if the DIC is present for person i, and $I_{ik} = 0$ if it is absent. Then the deprivation levels of person i, denoted D_i^*, is defined as

$$D_i^* = \sum_{k=1}^{K} \alpha_k^* I_{ik}, \qquad (2.14)$$

where $\alpha_k^* > 0$ is the weight attached to the kth DIC and is independent of the person being considered. If the weights relevant to the personal DICs are defined as $\alpha_k^* = 1 - p_k$, where p_k represents the frequency with which condition k occurs, then the α_k^* embodies the notion of "relative deprivation," i.e., the *smaller* the frequency with which a particular DIC is experienced, the *greater* the weight attached to it when it is experienced. The use of such weights echoes the work of Desai and Shah (1988) who, in a re-examination of Townsend's (1979) original data, essentially argued that to be deprived of something that almost everyone has is more important than to be deprived of something that few people possess.

It is only by accident that the weights, α_k^*, will sum to unity. They may, however be normalized by defining $\alpha_k = \alpha_k^*/\Omega$, where $\Omega = \sum_{k=1}^{K} \alpha_k^*$. Under this normalization, the deprivation level of person i may be defined as

$$D_i = D_i^*\Omega^{-1} = \left(\sum_{k=1}^{K} \alpha_k^* I_{ik} \right)\Omega^{-1} = \sum_{k=1}^{K} \alpha_k I_{ik}. \qquad (2.15)$$

Because D_i is simply a scalar transform of D_i^* the same ranking of individuals, in terms of their deprivation levels, will be obtained using D_i as using D_i^*. However, in terms of their normalized weights, the deprivation index D_i offers advantages of interpretation over D_i^* and the subsequent analysis will, therefore, be conducted in terms of this measure. Since, by definition, $\sum_{k=1}^{K} \alpha_k = 1$, Equation 2.15 implies that $0 \leq D_i \leq 1 : D_i = 0$ when none of the DICs are present, that is when $I_{ik} = 0$ for all $k = 1 \ldots K$ and $D_i = 1$ when all the DICs are present, that is when $I_{ik} = 1$ for all $k = 1 \ldots K$.

A major problem in constructing a deprivation index lies in deciding on the DICs that should enter its construction. Reviewing the literature on the construction of deprivation indices, Nolan and

Whelan (1996) pointed out that the role of tastes presented a major problem. If observed differences in living patterns could largely be ascribed to preferences rather than to resources then the absence of particular items could not be taken as an indicator of want. For example, Piachaud (1987) has highlighted the considerable variation in the deprivation scores of households at similar income levels. Even if one could separate preferences from needs, the importance of the researcher in choosing the items that enter the deprivation index is seen as a further problem. Then there is the question of whether deprivation should be measured solely by reference to an individual's own circumstance or, also, by reference to his or her wider social and geographical environment. Borooah and Carcach (1997), for example, drew attention to the importance of neighborhood-quality in determining the degree to which people were afraid of crime. Lastly, if one could arrive at a satisfactory set of indicators, there is the issue of how these are to be weighted in the construction of the overall index. Overarching these problems are the constraints imposed by the data; one can only construct a deprivation index from the data that are available, not from data that one might wish had been available. The deprivation index that I constructed was based upon data from the 1991 Census which gave information on the living circumstances of 13,164 individuals living in the region.[17] These data tell us, for example, whether, at the time of the Census, a person:

- Lived in a household in which none of the members normally had the use of a car or a van;
- Lived in a household in which none of the members were earners;
- Lived in a household in which the members did not have exclusive use of an inside toilet;
- Lived in a house without any central heating;
- Lived in a house without a public supply of water piped into the house;
- Lived in a house not connected to a public sewer;
- Lived in a house that represented nonpermanent accommodation;
- Lived in a house for which the ratio of the number of residents to the number of rooms was greater than 1;
- Suffered from a long-term illness, health problem or handicap which limited his or her daily activities or the work he or she could do.

Using this information, separate deprivation indices were constructed for retired and nonretired persons. Both indices were built around the same set of DICs, set out above, but the weights attached to these DICs differed according to whether the person was retired or not. As mentioned above, the weight associated with a DIC reflected the frequency with which the relevant DIC was experienced—the lower the frequency, the greater the weight. The frequencies with which several of the DICs were experienced were, however, considerably different for the retired and nonretired parts of the sample, and the use of different weights in constructing deprivation indices for retired and nonretired persons reflected this difference. Depending on the calculated value of his or her deprivation index, each person was then assigned to one of three deprivation levels—not deprived, mildly deprived, and severely deprived[18]—and associated with each outcome ($j = 1, 2, 3$) and each person ($i = 1, \ldots, N$), was a value of Y_i, an ordinal dependent variable, such that:

- $Y_i = 1$ if the person was not deprived
- $Y_i = 2$ if the person was mildly deprived
- $Y_i = 3$ if the person was severely deprived

It should be emphasized that it is these data, on the values of Y_i, that would typically be available to a researcher for analysis. Of the total sample of 13,164 persons, 45.9% (6,042 persons) were classed as being not deprived, 34.9% (4,594 persons) were classed as being mildly deprived, and 19.2% (2,528 persons) were classed as being severely deprived.

Equation Specification

The determining variables used to "explain" a person's deprivation level were:

- AGE_i in years, normalized by setting $AGE_i = 0$ for persons who were 16 years old;[19]
- $HIGHED_i = 1$, if the person had first, or higher, degree qualifications of UK standard; $HIGHED_i = 0$, otherwise;
- $MIDED_i = 1$, if the person had post-A level, but less than degree, qualifications[20]; $MIDED_i = 0$, otherwise;
- $RET_i = 1$, if the person was retired; $RET_i = 0$, otherwise;
- $INAC_i = 1$, if the person was economically inactive; $INAC_i = 0$, otherwise;

- $UE_i = 1$, if the person was unemployed; $UE_i = 0$, otherwise;
- $HNUM_i = 1$, if the number of persons in the household were six or more; $HNUM_i = 0$, otherwise;
- $SNPAR_i = 1$, if the person was a single parent; $SNPAR_i = 0$, otherwise;
- $AREA_{ai} = 1$, if person was resident in area a of Northern Ireland; $AREA_{ai} = 0$, otherwise. There were 10 such areas[21] identified in the 1991 Census for Northern Ireland.

In addition to these variables, it was possible that the level of deprivation might depend on the sex of a person and, because in Northern Ireland Catholics are a relatively disadvantaged group, also upon his or her religion. (Of 13,164 persons in the sample: 7,243 were men and 5,921 were women; 4,364 were Catholics and 8,800 were Protestants[22]). To account for this, two other variables were considered:

- $SEX_i = 1$, if the person was female, $SEX_i = 0$ otherwise;
- $CT_i = 1$, if the person was Catholic, $CT_i = 0$, otherwise.

Consequently, in the context of this application Equation 2.1, was specified as

$$D_i = \beta_1 + \beta_2 * SEX_i + \beta_3 * CT_i + \beta_4 * AGE_i$$
$$+ \beta_5 * AGE_i^2 + \beta_6 * HIGHED_i + \beta_7 * MIDED_i$$
$$+ \beta_8 * RET_i + \beta_9 * INAC_i + \beta_{10} * UE_i + \beta_{11} * HNUM_i$$
$$+ \beta_{12} * SNPAR_i + \beta_{13} * AREA_{2i} \cdots + \beta_{21} * AREA_{10i} + \varepsilon_i$$
$$= Z_i + \varepsilon_i. \tag{2.16}$$

The squared value of the age variable (AGE_i^2 above) introduces a nonlinearity to the age effect: the marginal effect of an increase in age upon D_i depends upon the age from which the increase takes place. If $\beta_4 < 0$ and $\beta_5 > 0$, then increasing a person's age reduces his or her deprivation score, but this effect is smaller the older the person is.[23]

The Equation Statistics

The estimated parameters $\hat{\beta}_k$ and $\hat{\delta}_1$ and $\hat{\delta}_2$ maximize the likelihood of observing the sample in which $N_1 = 6042$, $N_2 = 4594$, and

$N_3 = 2528$ (see Equation 2.4). These estimates are shown in Table 2.1 for the ordered logit model and in Table 2.2 for the ordered probit model. The z-ratios in Tables 2.1 and 2.2 are the ratios of the estimated coefficients to their estimated standard errors: the z-ratios are (asymptotically) distributed as $N(0, 1)$ under the null hypothesis that the associated coefficients are zero.[24]

Greene (2000, p. 831–833) has a number of suggestions for measuring the "goodness-of-fit" of equations with discrete dependent variables. He suggests that, at a minimum, one should report the maximized value of the log-likelihood function. The values of L_1 are the maximized log-likelihood values shown at the head of the tables (respectively, -12423.56 and -12426.25). Since the hypothesis that all the slopes in the model are zero is often interesting, the results of comparing the "full" model with an "intercept only" model should also be reported. The χ^2 values at the head of Tables 2.1 and 2.2 (respectively, 2571.98 and 2566.6) are defined as $2(L_1 - L_0)$, where L_0 is the value of the log-likelihood function when the only explanatory variable was the constant term and L_1, as observed earlier, is the value of the log-likelihood function when all the explanatory variables were included; the degrees of freedom are equal the number of slope coefficients estimated. These χ^2 values decisively reject the null hypothesis that the model did not have greater explanatory power than an "intercept only" model.[25]

The "pseudo-R^2" is defined as $1 - L_1/L_0$ and is due to McFadden (1973). This is bounded from below by 0 and from above by 1. A 0 value corresponds to all the slope coefficients being zero and a value of 1 corresponds to perfect prediction (that is, to $L_1 = 0$). Unfortunately, as Greene (2000) notes, the values between 0 and 1 have no natural interpretation, though it has been suggested that the pseudo-R^2 value increases as the fit of the model improves. Other measures have been suggested. Ben-Akiva and Lerman (1985) and Kay and Little (1986) suggested a fit measure which measured the average probability of correct prediction by the prediction rule. Cramer (1999) suggested a measure that corrected for the failure of Ben-Akiva/Lerman measure to take into account the fact that, in unbalanced samples, the less frequent outcome will usually be predicted very badly. In their survey of pseudo-R^2 measures, Veall and Zimmerman (1996) argued that in models of the multinomial probit or logit type, only the McFadden (1973) measure "seemed worthwhile."

TABLE 2.1
Ordered Logit on Deprivation in Northern Ireland

Ordered Logit Estimates; Log Likelihood = −12423.562; Number of obs = 13164; LR $\chi^2(20)$ = 2571.98; Prob > χ^2 = 0.0000; Pseudo R^2 = 0.0938

y	Coefficient	Standard Error	z	P > |z|	[95% Conf. Interval]	
sex	−0.1649064	0.0360722	−4.572	0.000	−0.2356066	−0.0942061
ct	0.1829855	0.0393496	4.650	0.000	0.1058617	0.2601092
age	−0.0253924	0.0049969	−5.082	0.000	−0.0351862	−0.0155987
age2	0.0008173	0.0000946	8.638	0.000	0.0006319	0.0010028
ret	0.6264438	0.0875236	7.157	0.000	0.4549007	0.797987
inac	1.297307	0.0765827	16.940	0.000	1.147207	1.447406
ue	1.460677	0.0619881	23.564	0.000	1.339182	1.582171
highed	−0.9081216	0.0683311	−13.290	0.000	−1.042048	−0.774195
mided	−0.6692574	0.1230722	−5.438	0.000	−0.9104745	−0.4280403
hnum	0.9214472	0.0551311	16.714	0.000	0.8133923	1.029502
snpar	0.305537	0.0637575	4.792	0.000	0.1805746	0.4304994
ard	−0.4497501	0.0642359	−7.002	0.000	−0.5756502	−0.3238499
dwn	0.0492598	0.0680278	0.724	0.469	−0.0840723	0.1825919
crk	−0.3161319	0.0706348	−4.476	0.000	−0.4545735	−0.1776903
ant	0.3472878	0.0717874	4.838	0.000	0.206587	0.4879887
col	0.4384781	0.0715933	6.125	0.000	0.2981578	0.5787985
arm	0.3824736	0.0740609	5.164	0.000	0.237317	0.5276303
ban	0.2364069	0.066677	3.546	0.000	0.1057223	0.3670914
dry	0.0791624	0.0787678	1.005	0.315	−0.0752198	0.2335445
frm	0.6350174	0.0747171	8.499	0.000	0.4885745	0.7814602
_cut1	0.1804476	0.074681			(Ancillary parameters)	
_cut2	2.058464	0.0772157				

y	Probability	Observed
1	Pr($xb + u < $ _cut1)	0.4591
2	Pr(_cut1 $< xb + u < $ _cut2)	0.3493
3	Pr(_cut2 $< xb + u$)	0.1917

TABLE 2.2

Ordered Probit on Deprivation in Northern Ireland

Ordered Probit Estimates; Log Likelihood = -12426.255; Number of obs = 13164; LR $\chi^2(20)$ = 2566.60; Prob > χ^2 = 0.0000; Pseudo R^2 = 0.0936

y	Coefficient	Standard Error	z	P > |z|	[95% Conf. Interval]	
sex	-0.1003127	0.021542	-4.657	0.000	-0.1425343	-0.0580911
ct	0.112619	0.0235271	4.787	0.000	0.0665067	0.1587314
age	-0.0154206	0.002965	-5.201	0.000	-0.021232	-0.0096093
age2	0.0004965	0.000056	8.872	0.000	0.0003868	0.0006062
ret	0.3679154	0.0528714	6.959	0.000	0.2642894	0.4715415
inac	0.7680422	0.0445939	17.223	0.000	0.6806398	0.8554446
ue	0.8555121	0.0357888	23.904	0.000	0.7853673	0.925657
highed	-0.5562051	0.0401161	-13.865	0.000	-0.6348313	-0.4775789
mided	-0.4048568	0.0721096	-5.614	0.000	-0.546189	-0.2635246
hnum	0.5459127	0.0328011	16.643	0.000	0.4816237	0.6102017
snpar	0.1748777	0.0379915	4.603	0.000	0.1004157	0.2493396
ard	-0.2729771	0.03806	-7.172	0.000	-0.3475734	-0.1983808
dwn	0.0216987	0.0403927	0.537	0.591	-0.0574696	0.100867
crk	-0.1831732	0.0417875	-4.383	0.000	-0.2650752	-0.1012712
ant	0.2026288	0.0428712	4.726	0.000	0.1186028	0.2866547
col	0.2518903	0.0428189	5.883	0.000	0.1679668	0.3358137
arm	0.2122576	0.0444114	4.779	0.000	0.1252129	0.2993024
ban	0.129325	0.0399411	3.238	0.001	0.0510418	0.2076082
dry	0.0360175	0.0471155	0.764	0.445	-0.0563271	0.1283622
frm	0.3626456	0.0447755	8.099	0.000	0.2748871	0.450404
_cut1	0.1009536	0.0444785			(Ancillary parameters)	
_cut2	1.212453	0.0453671				

y	Probability	Observed
1	Pr($xb + u < $ _cut1)	0.4591
2	Pr(_cut1 $< xb + u <$ _cut2)	0.3493
3	Pr(_cut2 $< xb + u$)	0.1917

An alternative to "point" measures of goodness-of-fit might be to assess the predictive ability of the model. Such assessments are routine in models of binary choice where the hits ($Y_i = 1$) and misses ($Y_i = 0$) predicted by the model *on the basis of a prediction rule* (say, $Y_i = 1$ if $\hat{p}_i > 0.5$, $Y_i = 0$, otherwise) are compared to the actual hits and misses. This procedure could be extended to multiple outcome models where the predictions could be based on a rule whereby $Y_i = m$, if $\hat{p}_{im} = \text{Max}_j (\hat{p}_{ij})$. These predictions could then be compared to a "naive" model that predicted all cases to be in the modal category of the dependent variable, and the percentage reduction in error in moving from the naive to the full model could be computed. This approach is, however, not without pitfalls. First, unlike the case of the linear regression model, where the coefficients are chosen to maximize R^2, in discrete choice models the coefficient estimates do not maximize *any* goodness-of-fit measure. So to assess the model on the basis of goodness-of-fit, however measured, may be misleading. Second, the predictions are critically dependent on the prediction rule adopted, and the adopted rule may turn out to be quite inappropriate to the needs of the sample. For example, in a binary model, if the sample is unbalanced—that is, has many more 1s than 0s—then a rule that the model should predict the outcome for which estimated probability is greatest might never predict a 1 (or a 0).

The Estimates

The coefficients in Tables 2.1 and 2.2 are all significantly different from zero except for the two area coefficients associated with living in County Down and in County Derry (variables *dwn* and *dry* in the tables). The items under the column "$P > |z|$" in Table 2.1 show that under the null hypothesis that the coefficients on *dwn* and *dry* were zero, there was, respectively, a 46.9% and a 31.5% chance of observing a value in the distribution tails *beyond* the observed values, 0.0492598 and 0.0791624. The items under the column "$P > |z|$" in Table 2.2 show these chances as 59.1% and 44.5%, respectively. As regards the other coefficients, the items under the column "$P > |z|$" in Tables 2.1 and 2.2 show that, under the null hypothesis that they were zero, there was a "zero" chance of observing a value in the distribution tails *beyond* the observed values. The items under the column "95% Conf. Interval" show the limits *outside* which the estimates must lie if their associated coefficients are to be regarded as different from zero at the 5% level of significance. In both Tables 2.1 and 2.2, the estimates on *dwn* and *dry* coefficients lay within their 95% limits, while the other coefficient estimates lay outside their limits.

In interpreting the individual coefficients the importance of the ceteris paribus ("other things equal") clause must be emphasized. The estimate on the variable *ct* was positive, indicating that ceteris paribus Catholics had a higher probability of being severely deprived and a lower probability of being not deprived than Protestants. This is not to say that every Catholic had a higher probability of being severely deprived than every Protestant. Rather the correct interpretation is that, given two persons who were similar in respect of every characteristic *except* religion, the person who was Catholic was more likely to be severely deprived and less likely to be not deprived than the person who was Protestant. The estimate on the variable *sex* was negative, indicating that ceteris paribus women had a lower probability of being severely deprived and a higher probability of being not deprived than men. The other coefficients carry a similar interpretation. For example, the negative estimates on the variables *highed* and *mided*, indicate that ceteris paribus persons with educational qualifications had a lower probability of being severely deprived and a higher probability of being not deprived than persons without educational qualifications. As the discussion of the previous discussion indicated, the signs of the coefficient estimates allow only the direction of change in the probabilities of the extreme outcomes, following a change in the value of the associated variable, to be predicted. The direction of change in the probabilities of the intermediate outcomes cannot be inferred. For example, from an inspection of the estimates we cannot say whether the probability of a woman being mildly deprived is larger or smaller than that of a man.

The Cutoff Points

The estimated cutoff points are shown below the estimates. A display at the bottom of Tables 2.1 and 2.2 shows how the probabilities for the categories were computed from the fitted equation.[26] This display was generated by including the "table" option in the syntax of the oprobit and ologit commands set out in the program listing in Chapter 4. The assumption behind ordinal regression is that the observed categories represent crude but correctly ordered differences on an underlying continuous scale. The probabilities of belonging to these categories are defined in terms of the probabilities of the values of an underlying latent variable crossing particular thresholds, where these thresholds are established by the values of the cutoff points.

In the ordered logit points the two cutoffs were estimated[27] as _cut1 = 0.1804476 and _cut2 = 2.058464 where, in the notation of Equations 2.5a to 2.5c, $\hat{\delta}_1$ = _cut1 and $\hat{\delta}_2$ = _cut2. Consequently, in the ordered logit model, the probability of a person being not deprived was $\Pr(\widehat{Z}_i + \varepsilon_i \leq 0.1804476)$, being mildly deprived was $\Pr(0.1804476 \leq \widehat{Z}_i + \varepsilon_i \leq 2.058464)$, and being severely deprived was $\Pr(\widehat{Z}_i + \varepsilon_i \geq 2.058464)$. In the ordered probit model the two cutoffs were estimated as _cut1 = 0.1009536 and _cut2 = 1.212453. In the ordered probit model, the probability of a person being not deprived was $\Pr(\widehat{Z}_i + \varepsilon_i \leq 0.1009536)$, being mildly deprived was $\Pr(0.1009536 \leq \widehat{Z}_i + \varepsilon_i \leq 1.212453)$, and being severely deprived was $\Pr(\widehat{Z}_i + \varepsilon_i \geq 1.212453)$. The reason the estimated cutoff points were different in the two models is that the slope coefficient estimates were also substantially different. However, the conjunction of the estimated slope and cutoff coefficients in each model meant that the predictions from the two models were very similar.

The Predicted Probabilities: Calculation From Individuals

Using the estimated $\widehat{Z}_i$—which, remembering that $\widehat{Z}_i = \sum_{k=1}^{K} \hat{\beta}_k X_{ik}$, are computed using the estimates shown in Table 2.1, in conjunction with the values of the determining variables for every individual— STATA will predict, for *each* of the 13,164 individuals in the sample, the probabilities of belonging to the three different deprivation levels, by computing the $\hat{p}_{i1}$, $\hat{p}_{i2}$, and $\hat{p}_{i3}$ of, respectively, Equations 2.5a, 2.5b, and 2.5c. Table 2.3 shows, using the logistic distribution for the error term ε_i, these calculations for the first twenty five persons in the sample. We see from Table 2.3 that, given his or her circumstances, person 17 had a very low probability of being severely deprived (4%) and a high probability of being not deprived (78%). On the other hand, person 2's circumstances meant that he or she had a very high probability of being severely deprived (69%) and a very low probability of being not deprived (6%). It is instructive to compare the predicted probabilities from the logit model with the predicted probabilities obtained from assuming that the ε_i were normally distributed. These are shown in Table 2.4. A comparison of Tables 2.3 and 2.4 shows that, notwithstanding the differences in coefficient estimates between the logit and probit models, the predicted

TABLE 2.3

Ordered Logit Calculated Probabilities of Being at Different
Deprivation Levels: First 25 Persons in the Sample

p_{num}	p1	p2	p3
1	0.2468799	0.4350583	0.3180617
2	0.0635542	0.2438714	0.6925744
3	0.1456977	0.3815914	0.472711
4	0.221723	0.4290327	0.3492444
5	0.1984064	0.4197503	0.3818433
6	0.0774309	0.2769676	0.6456016
7	0.3072922	0.4363922	0.2563156
8	0.6020886	0.3061397	0.0917717
9	0.6067917	0.303063	0.0901453
10	0.5668221	0.3285579	0.10462
11	0.3003936	0.4370245	0.2625819
12	0.4497833	0.3926532	0.1575635
13	0.1598765	0.3946234	0.4455001
14	0.5337384	0.348435	0.1178265
15	0.6265731	0.2899149	0.0835119
16	0.5928456	0.3121286	0.0950258
17	0.7833934	0.1760466	0.04056
18	0.5323462	0.3492447	0.1184092
19	0.5169058	0.3580674	0.1250268
20	0.6024157	0.3059263	0.091658
21	0.5469929	0.3406158	0.1123913
22	0.5918246	0.3127853	0.0953901
23	0.7823876	0.1768215	0.0407909
24	0.5389312	0.3453955	0.1156733
25	0.4514354	0.3918849	0.1566797

p_{num} = person number
p1 = probability of being "not deprived"
p2 = probability of being "mildly deprived"
p3 = probability of being "severely deprived"

probabilities are very similar. For everyone of the 25 persons listed, the predicted probabilities of being at the three deprivation levels are not very different under the logit and probit formulations. In that sense, it did not matter which model was used; the predicted outcomes were very similar.

The individual probabilities can be used to generate sample statistics of deprivation levels. These are shown in Table 2.5 for the logit model and in Table 2.6 for the probit model. The mean values of $\hat{p}_{i1}$,

TABLE 2.4

Ordered Probit Calculated Probabilities of Being at Different
Deprivation Levels: First 25 Persons in the Sample

P_{num}	$p1$	$p2$	$p3$
1	0.2564767	0.4197788	0.3237445
2	0.0561468	0.2607234	0.6831297
3	0.148693	0.3789889	0.4723181
4	0.2297609	0.4152423	0.3549968
5	0.2077654	0.4091158	0.3831188
6	0.0723101	0.2918726	0.6358173
7	0.3164209	0.4204631	0.263116
8	0.5960203	0.3162006	0.0877791
9	0.6006402	0.3134694	0.0858905
10	0.5609338	0.3361012	0.102965
11	0.3078035	0.4210712	0.2711253
12	0.4515077	0.3873188	0.1611736
13	0.1692549	0.3920921	0.438653
14	0.5286788	0.3530054	0.1183157
15	0.6201291	0.3016761	0.0781948
16	0.5868787	0.3215305	0.0915909
17	0.7813467	0.189157	0.0294963
18	0.5321902	0.3512332	0.1165766
19	0.5122991	0.3610432	0.1266577
20	0.5962939	0.3160395	0.0876666
21	0.5415704	0.3464162	0.1120134
22	0.5858997	0.3220953	0.092005
23	0.7803024	0.1899634	0.0297342
24	0.5342544	0.3501835	0.1155621
25	0.449369	0.3881331	0.1624979

P_{num} = person number
$p1$ = probability of being "not deprived"
$p2$ = probability of being "mildly deprived"
$p3$ = probability of being "severely deprived"

$\hat{p}_{i2}$, and $\hat{p}_{i3}$, as shown in Tables 2.5 and 2.6 (referred to as $\bar{p}_1$, $\bar{p}_2$, and $\bar{p}_3$) were slightly different between the logit and probit models. However, *after rounding* , these differences disappeared and the estimates were respectively: 46%, 35% and 19%. Under the logit model, the mean probabilities, as calculated above, are equal to the sample proportions in the three categories of deprivation.[28] This is a property of the ordered logit model. The mean probabilities under the probit model are close, but not equal, to the sample proportions and this too is regularly observed in practice. The mean probabilities as set out in Tables 2.5 and 2.6 sum to precise unity. This is because the individual probability estimates (from which the mean is computed) sum to unity.

The median (50th percentile) values of $\hat{p}_{i1}$, $\hat{p}_{i2}$ and $\hat{p}_{i3}$, as shown in Tables 2.5 and 2.6, were also slightly different between the logit and probit models though, again, these differences disappeared after rounding, yielding estimates of the median probabilities: 48%, 37% and 14%, respectively. It is not necessary that the median probabilities sum to unity for any of the two models and, indeed, in this application they fall short of unity. The reason for this is that while, for *any one* person the estimates must sum to unity, the median values of $\hat{p}_{i1}$, $\hat{p}_{i2}$ and $\hat{p}_{i3}$ need not relate to the *same* person. For example, if the values of $\hat{p}_{i1}$, $\hat{p}_{i2}$ and $\hat{p}_{i3}$ ($i = 1, \ldots, 13, 164$) were arranged in ascending or descending order of magnitude, the person at the 50th percentile for $\hat{p}_{i1}$ may be different from the person at the 50th percentile for $\hat{p}_{i3}$. On the logit estimates (Table 2.5), the probability of being not deprived ($\hat{p}_{i1}$) at the lowest percentile was 6.7% and, within this class, the lowest estimated value was 1.9%. At the other end of the scale, the value of $\hat{p}_{i1}$ at the highest percentile was 83.1% and, within this class, the highest estimated value was 87%.

Predicted Probabilities: Calculation at the Mean

In the previous subsection the estimates of the mean probabilities were computed from the estimates of the *individual* probabilities, $\hat{p}_{i1}$, $\hat{p}_{i2}$, and $\hat{p}_{i3}$ ($i = 1, \ldots, 13, 164$). Under the logit model the mean probabilities were exactly equal to the sample proportions, and under the probit model they were close to the sample proportions. However, there is an alternative way of calculating mean probabilities (see Greene, 2000, p. 879) and this leads to a different outcome from that set out above. Let $\overline{X}_k = \sum_{i=1}^{N} X_{ik}/N$ be the mean value of the kth determining variable and let

$$\overline{Z} = \sum_{i=1}^{N} \hat{Z}_i/N = \sum_{i=1}^{N}\left(\sum_{k=1}^{K} \hat{\beta}_k X_{ik} \right)\bigg/ N = \sum_{k} \hat{\beta}_k \sum_{i} X_{ik}/N$$

$$= \sum_{k} \hat{\beta}_k \overline{X}_k \qquad (2.17)$$

TABLE 2.5
Sample Statistics of Deprivation Levels: Logit Model

			Pr ($y = 1$)	
	Percentiles	Smallest		
1%	0.0675202	0.0185615		
5%	0.1466481	0.0198162		
10%	0.1875675	0.0300967	Obs	13164
25%	0.303902	0.0347162	Sum of Wgt.	13164
50%	0.4843619		Mean	0.4592781
		Largest	Standard Deviation	0.1889648
75%	0.5993176	0.8697239		
90%	0.6954355	0.8697425	Variance	0.0357077
95%	0.7368268	0.8699151	Skewness	−0.1866478
99%	0.8308569	0.8699151	Kurtosis	2.155329
			Pr ($y = 2$)	
	Percentiles	Smallest		
1%	0.1389147	0.0915195		
5%	0.2075179	0.0969696		
10%	0.239251	0.1077327	Obs	13164
25%	0.2989144	0.1077327	Sum of Wgt.	13164
50%	0.3674155		Mean	0.3493639
		Largest	Standard Deviation	0.076207
75%	0.4150781	0.4377985		
90%	0.4339717	0.4377985	Variance	0.0058075
95%	0.4366746	0.4377985	Skewness	−0.8337451
99%	0.4377566	0.4377985	Kurtosis	2.917122
			Pr ($y = 3$)	
	Percentiles	Smallest		
1%	0.0301859	0.0223522		
5%	0.0517813	0.0223522		
10%	0.0627571	0.0223856	Obs	13164
25%	0.0927392	0.0223891	Sum of Wgt.	13164
50%	0.1399815		Mean	0.191358
		Largest	Standard Deviaition	0.1405856
75%	0.2593728	0.8095671		
90%	0.3984033	0.8312855	Variance	0.0197643
95%	0.4708132	0.8832142	Skewness	1.451671
99%	0.6786129	0.889919	Kurtosis	4.969932

TABLE 2.6
Sample Statistics of Deprivation Levels: Probit Model

			Pr $(y = 1)$	
	Percentiles	Smallest		
1%	0.0610129	0.0089802		
5%	0.1495682	0.0095785		
10%	0.1936998	0.0190097	Obs	13164
25%	0.311691	0.0255515	Sum of Wgt.	13164
50%	0.4870896		Mean	0.4602124
		Largest	Standard Deviation	0.1855043
75%	0.5928672	0.8747257		
90%	0.6891779	0.8747432	Variance	0.0344118
95%	0.7340055	0.8749362	Skewness	−0.212917
99%	0.8299778	0.8749362	Kurtosis	2.250704
			Pr $(y = 2)$	
	Percentiles	Smallest		
1%	0.1500103	0.0957713		
5%	0.2209043	0.09959		
10%	0.2547602	0.1132009	Obs	13164
25%	0.3088254	0.1132009	Sum of Wgt.	13164
50%	0.3662931		Mean	0.3488479
		Largest	Standard Deviation	0.0663654
75%	0.4044376	0.421618		
90%	0.418923	0.421618	Variance	0.0044044
95%	0.4207648	0.421618	Skewness	−1.039971
99%	0.421588	0.421618	Kurtosis	3.528536
			Pr $(y = 3)$	
	Percentiles	Smallest		
1%	0.0194342	0.011863		
5%	0.0412402	0.011863		
10%	0.0542446	0.011892	Obs	13164
25%	0.0890826	0.0118946	Sum of Wgt.	13164
50%	0.1402645		Mean	0.1909397
		Largest	Standard Deviation	0.1432815
75%	0.2674853	0.7992974		
90%	0.4023936	0.8322631	Variance	0.0205296
95%	0.4708187	0.8908314	Skewness	1.315678
99%	0.668156	0.8952485	Kurtosis	4.483449

be the mean value of the $\widehat{Z}_i$. Then compute $\hat{p}_1$, $\hat{p}_2$, and $\hat{p}_3$ as

$$\hat{p}_1 = \Pr(\varepsilon_i \leq \hat{\delta}_1 - \overline{Z}) \qquad (2.18a)$$

$$\hat{p}_2 = \Pr(\hat{\delta}_1 - \overline{Z} < \varepsilon_i \leq \hat{\delta}_2 - \overline{Z}) \qquad (2.18b)$$

$$\hat{p}_3 = \Pr(\varepsilon_i \geq \hat{\delta}_2 - \overline{Z}). \qquad (2.18c)$$

Then, in general: $\bar{p}_1 \neq \hat{p}_1$, $\bar{p}_2 \neq \hat{p}_2$, and $\bar{p}_3 \neq \hat{p}_3$. This is not surprising because while both the $\bar{p}_j$ and the $\hat{p}_j$ ($j = 1, 2, 3$), purport to measure the overall probability of being at the different levels of deprivation, they are computed very differently. The $\bar{p}_j$ are computed as the mean of the estimated individual probabilities, $\hat{p}_{ij}$, and the $\hat{p}_{ij}$ are computed from the values of the determining variables for the individual in conjunction with the estimated cutoff points. On the other hand, the $\hat{p}_j$, which bypass the individual probabilities, are computed instead from the mean value, $\overline{Z}$, of the individual Z_i (or equivalently, calculated using the mean values, $\overline{X}_k$ of the determining variables, X_{ik}) in conjunction with the estimated cutoff points. Tables 2.7 and 2.8 compare the values of $\bar{p}_j$ and $\hat{p}_j$ ($j = 1, 2, 3$) for the logit and probit models, respectively.

Calculation of Marginal Probabilities: Continuous Variables

The only continuous variable in the model, as specified in Equation 2.16, are AGE_i and AGE_i^2. In order to compute the effect of an increase of one year in the age of a person on the probabilities of his or her being at the three different deprivation levels, one needs to evaluate the appropriate density functions at the relevant points and multiply these by the coefficient estimates associated

TABLE 2.7
Overall Probability of Being at Different Deprivation Levels:
Logit Model

Probability of Being:	CALCULATED AS: Mean of Individual Probabilities: $\bar{p}$	CALCULATED FROM: Mean of Determining Variables: $\hat{p}$
Not deprived	0.459	0.447
Mildly deprived	0.349	0.394
Severely deprived	0.191	0.159

TABLE 2.8
Overall Probability of Being at Different Deprivation Levels: Probit Model

Probability of Being:	CALCULATED AS: Mean of Individual Probabilities: $\bar{p}$	CALCULATED FROM: Mean of Determining Variables: $\hat{p}$
Not deprived	0.460	0.452
Mildly deprived	0.349	0.387
Severely deprived	0.191	0.161

with the two variables (see Equations 2.12a to 2.12c and 2.13a to 2.13c). This can be done in either of two ways:

1. For each individual in the sample, evaluate the marginal effects at the relevant points for that individual. These relevant points for individual i ($i = 1 \ldots 13{,}164$) are: $\hat{\delta}_1 - \widehat{Z}_i$ for the marginal effect on $\Pr(Y_i = 1)$; $\hat{\delta}_2 - \widehat{Z}_i$ and $\hat{\delta}_1 - \widehat{Z}_i$ for the marginal effect on $\Pr(Y_i = 2)$; and $\hat{\delta}_2 - \widehat{Z}_i$ for the marginal effect on $\Pr(Y_i = 3)$. Multiply these evaluations by the coefficient estimate $(\hat{\beta}_4)$ on AGE_i and also multiply these evaluations by the coefficient estimate $(\hat{\beta}_5)$ on AGE_i^2. Add these two results to obtain, for each individual, the marginal effect on his or her probabilities of a small (one year) change in age. The mean of these individual effects then yields the *average* effect on the probabilities (of being at the different deprivation levels) of increasing the age of *every person in the sample* by one year.

2. Compute the mean value of $\widehat{Z}_i$ over all the persons in the sample (see Equation 2.17). If this is denoted $\overline{Z}$, evaluate the marginal effects at the relevant points: $\hat{\delta}_1 - \overline{Z}$ for the marginal effect on $\Pr(Y_i = 1)$; $\hat{\delta}_2 - \overline{Z}$ and

TABLE 2.9
Marginal Effect of an Additional Year of Age on the Probability of Being at Different Deprivation Levels: Logit Model

Probability of Being:	CALCULATED AS: Mean of Individual Marginal Effects	CALCULATED FROM: Mean of Determining Variables
Not deprived	−0.00523	−0.00607
Mildly deprived	0.00191	0.00279
Severely deprived	0.00332	0.00329

$\hat{\delta}_1 - \overline{Z}$ for the marginal effect on $\Pr(Y_i = 2)$; and $\hat{\delta}_2 - \overline{Z}$ for the marginal effect on $\Pr(Y_i = 2)$. Multiply these evaluations by the coefficient estimate on AGE_i and also multiply these evaluations by the coefficient estimate on AGE_i^2. Add these two results to obtain the effect on the probabilities (of being at the different deprivation levels) of increasing the *average age of the persons in the sample* by one year.

Tables 2.9 and 2.10 shows the marginal effects of an increase in age, computed in the two different ways described above, for the logit and probit models.

Calculation of Marginal Probabilities: Dummy Variables

Earlier it was observed that the effects of a dummy variable should be analyzed by comparing the probabilities that result when the dummy variable takes one value with the probabilities that are the consequence of it taking the other value, the values of the other variables remaining unchanged between the two comparisons. This methodology is now used to analyze the effect of religion on the probabilities of being at different levels of deprivation by comparing the situation in which (for all i) $CT_i = 1$ (the religion dummy) with the situation in which (for all i) $CT_i = 0$. As with the case of continuous variables, this broad methodology can be implemented in either of two ways:

1. Suppose ceteris paribus that all the persons in the sample were Catholic so that $CT_i = 1$ ($i = 1 \ldots 13, 164$). Then for this scenario, $\widehat{Z}_i$ (the estimated value of Z_i) is computed from Equation 2.16, using the coefficient estimates $\hat{\beta}_k$ and with $CT_i = 1$ for all i. Call this estimated

TABLE 2.10

Marginal Effect of an Additional Year of Age on the Probability of Being at Different Deprivation Levels: Probit Model

Probability of Being:	CALCULATED AS: Mean of Individual Marginal Effects	CALCULATED AT: Mean of Determining Variables
Not deprived	−0.00524	−0.00591
Mildly deprived	0.00171	0.00226
Severely deprived	0.00353	0.00365

value $\widehat{Z}_i^C$. Let $\hat{p}_{ij}^C$ denote the (computed) probability of person i being at deprivation level j ($j = 1, 2, 3$) in this hypothetical situation, where these probabilities are computed from Equations 2.5a to 2.5c—or, if a probit model is used, from Equations 2.11a to 2.11c—with $\widehat{Z}_i = \widehat{Z}_i^C$. Now suppose ceteris paribus that all the persons in the sample were Protestant so that $CT_i = 0$ for all $i = 1 \dots 13, 164$ and let $\widehat{Z}_i^P$ represent the estimated value of Z_i. From Equation 2.16, $\widehat{Z}_i^C = \widehat{Z}_i^P + \hat{\beta}_3$. Let $\hat{p}_{ij}^P$ denote the (computed) probability of person i being at deprivation level j ($j = 1, 2, 3$) under these hypothetical circumstances. For any person i in the sample, the difference between the $\hat{p}_{ij}^C$ and $\hat{p}_{ij}^P$ is entirely due to religion: the $\hat{p}_{ij}^C$ are computed using $\widehat{Z}_i^C$—that is, with the Catholic variable "switched on"—and the $\hat{p}_{ij}^P$ are computed using $\widehat{Z}_i^P$, with the Catholic variable "switched off," without the value of any other variable being altered. If $\bar{p}_j^C = \sum_{i=1}^{N} \hat{p}_{ij}^C / N$ and $\bar{p}_j^P = \sum_{i=1}^{N} \hat{p}_{ij}^P / N$ are the respective means of the two sets of probability estimates, then the difference between them measures the effect of religion on the mean probability of being at different deprivation levels.

2. An alternative is to compare the probabilities that result when the variable CT_i takes its two different values across all the persons in the sample, with the values of the other variables, in each case, held at their sample means. Denote by $\overline{Z}^C$ and $\overline{Z}^P$ the values of $\overline{Z}$ when, respectively, $CT_i = 1$ and $CT_i = 0$ for all $i = 1 \dots 13, 164$ with, in each case, $X_{ik} = \overline{X}_k$ for the other variables. From Equation 2.16, $\overline{Z}^C = \overline{Z}^P + \hat{\beta}_3$. This exercise, in effect, constructs a "straw person" who, apart from religion, embodies the average qualities of the sample and who is Catholic in one scenario and Protestant in another. Then, using $\overline{Z}^C$ and then $\overline{Z}^P$ in Equations 2.5a to 2.5c, estimates of the three probabilities (of being not deprived, mildly deprived, and severely deprived) can be computed for the two hypothetical situations in which first this person is Catholic and second this person is Protestant. If these two sets of probabilities are denoted, respectively, $\hat{q}_j^C$ and $\hat{q}_j^P$ ($j = 1, 2, 3$) then their difference measures the effect of religion on the probability of being at different deprivation levels.

Table 2.11 shows the values of $\bar{p}_j^C$ and $\bar{p}_j^P$ and of $\hat{q}_j^C$ and $\hat{q}_j^P$ ($j = 1, 2, 3$) computed under the logit model, and Table 2.12 does the same for the probit model. Two features of these Tables are significant. First, for any of the two ways of computing marginal effects there was hardly any difference between the logit and probit probabilities. Second, for any one model there was considerable difference between the probabilities calculated in the two different ways. Notice that the probability of being severely deprived was significantly higher

TABLE 2.11

The Effect of Religion on the Probability of Being at Different
Deprivation Levels: Logit Model

Probability of Being:	CALCULATED AS: Mean of Individual Marginal Effects		CALCULATED AT: Mean of Determining Variables	
	CAT	PRT	CAT	PRT
Not deprived	0.432	0.472	0.417	0.462
Mildly deprived	0.360	0.346	0.407	0.387
Severely deprived	0.207	0.182	0.176	0.151

when the probabilities were computed as the mean of the individual probabilities than when they were computed from the average characteristics of the sample. This is not surprising. Being in a state of severe deprivation is the result of possessing "extreme" values of the deprivation determining variables. The influence of extreme values is dampened when the individual values are set equal to the sample averages. On the other hand, extreme values are allowed full play when the individual values are used in the probability calculations.

So which method is the appropriate one to use? The critical question is how the values of the other variables are to be held constant, when the dummy variable of interest takes its two different values. The second method, in effect, assigns to each individual the values of the sample means. But, of course, there is no sanctity to the mean. The common value assigned could be the median or any other value, and each of these different assignments would lead to a different out-

TABLE 2.12

The Effect of Religion on the Probability of Being at Different
Deprivation Levels: Probit Model

Probability of Being:	CALCULATED AS: Mean of Individual Marginal Effects		CALCULATED AT: Mean of Determining Variables	
	CAT	PRT	CAT	PRT
Not deprived	0.433	0.473	0.421	0.466
Mildly deprived	0.359	0.346	0.398	0.382
Severely deprived	0.208	0.191	0.181	0.152

come in terms of the calculated probabilities, $\hat{q}_j$. The first methodology does not suffer from this defect. Since the individual values are not interfered with, there would be a unique outcome in terms of the calculated individual probabilities, and therefore a unique outcome, under the two scenarios, in terms of the mean[29] probabilities, $\bar{p}_j$.

Estimation Over Subsamples:
Characteristics Versus Coefficients

Religion had an effect on the probabilities of being at different deprivation levels (Tables 2.11 and 2.12) because of the presence of β_3 in Equation 2.16. This means that, under the first methodology, $\hat{Z}_i^C \neq \hat{Z}_i^P$ and that, under the second, $\overline{Z}^C \neq \overline{Z}^P$. Consequently, the probabilities of being deprived were different when for each person in the sample $CT_i = 1$ than when $CT_i = 0$. But this leads one to consider the possibility that for *every variable* in Equation 2.16, the "Catholic" coefficients are different from the "Protestant" coefficients. In other words, Equation 2.16 should have been estimated allowing the Catholic and Protestant coefficients to be different from each other. There are two (almost equivalent) ways of implementing this. The first is to estimate a *single* equation but allow each Catholic coefficient to be different from the corresponding Protestant coefficient in the equation. The second is to estimate two *separate* equations on the Catholic and Protestant subsamples. In order to implement the first strategy, define the equation as

$$
\begin{aligned}
D_i = {} & \beta_1 + \gamma_1 * CT_i + \beta_2 * SEX_i \\
& + \gamma_2 * CT_i * SEX_i + \beta_3 * AGE_i \\
& + \gamma_3 * CT_i * AGE_i + \beta_4 * AGE_i^2 \\
& + \gamma_4 * CT_i * AGE_i^2 + \beta_5 * HIGHED_i \\
& + \gamma_5 * CT_i * HIGHED_i + \beta_6 * MIDED_i \\
& + \gamma_6 * CT_i * MIDED_i + \beta_7 * RET_i \\
& + \gamma_7 * CT_i * RET_i + \beta_8 * INAC_i \\
& + \gamma_8 * CT_i * INAC_i + \beta_9 * UE_i \\
& + \gamma_9 * CT_i * UE_i + \beta_{10} * HNUM_i
\end{aligned}
\tag{2.19}
$$

$$+ \gamma_{10} * CT_i * HNUM_i + \beta_{11} * SNPAR_i$$

$$+ \gamma_{11} * CT_i * SNPAR_i$$

$$+ \beta_{13} * AREA_{2i} \cdots + \beta_{21} * AREA_{10i}$$

$$+ \gamma_{13} * CT_i * AREA_{2i} \cdots + \gamma_{21} * CT_i * AREA_{10i} + \varepsilon_i$$

$$= Z_i + Z_i^C + \varepsilon_i.$$

By including "interaction variables" in the equation—that is, the original explanatory variables shown in Equation 2.16 *multiplied by the Catholic dummy variable*—the effect of a particular variable on deprivation is allowed to be different, depending upon whether the person concerned is Protestant or Catholic. In Equation 2.19, the β_k are the "Protestant" coefficients[30] and the γ_k—the coefficients attached to the interaction variables—represent the *additional* contribution to these coefficients resulting from being Catholic. The γ_k values measure, therefore, the strength of the "interaction effects." A test of $\gamma_k = 0$ is, therefore, a test of the null hypothesis that there is no interaction between the kth explanatory variable and the religion of a person. To put it differently, $\gamma_k = 0$ implies that there is no difference between the Protestant and Catholic coefficients on the kth explanatory variable. The big advantage of this single equation "integrated" approach, over the two separate equations approach, is that it makes it possible to easily test whether some variable had a differential impact on the two groups. Indeed, the single equation "integrated" approach is explored in some depth in the subsequent chapter. However, to give a flavor of the second approach, Equation 2.16 was estimated *separately* on the Catholic and Protestant subsamples. Tables 2.13 and 2.14 show, respectively, the results of estimating Equation 2.16 separately for Catholics and for Protestants, when the errors are assumed to be logistically distributed.[31]

In the equation estimates shown in Tables 2.13 and 2.14, the variables *dwn* and *dry* were dropped. This is because their coefficients were individually (on the z-score) and jointly (on a likelihood ratio test) not significantly different from zero. This is an important methodological point: when equations are used for prediction should they contain all the variables, even though some of the coefficients may not be significantly different from zero, or should they

TABLE 2.13
Ordered Logit on Deprivation in Northern Ireland: Protestant Subsample

Ordered Logit Estimates Log Likelihood = −8124.0888; Number of obs = 8800; LR $\chi^2(17)$ = 1527.75; Prob > χ^2 = 0.0000; Pseudo R^2 = 0.0859

y	Coefficient	Standard Error	z	P > |z|	[95% Conf. Interval]	
sex	−0.2293014	0.0444134	−5.163	0.000	−0.31635	−0.1422527
age	−0.0275073	0.0060613	−4.538	0.000	−0.0393873	−0.0156273
age2	0.0008768	0.0001131	7.754	0.000	0.0006551	0.0010984
ret	0.6275465	0.1008167	6.225	0.000	0.4299493	0.8251437
inac	1.179892	0.0954488	12.362	0.000	0.9928158	1.366968
ue	1.556987	0.0848854	18.342	0.000	1.390615	1.723359
highed	−0.8085732	0.085388	−9.469	0.000	−0.9759306	−0.6412159
mided	−0.6320822	0.1490549	−4.241	0.000	−0.9242244	−0.33994
hnum	0.95203	0.08401	11.332	0.000	0.7873735	1.116686
snpar	0.3647026	0.0845436	4.314	0.000	0.1990001	0.5304051
ard	−0.4509469	0.063093	−7.147	0.000	−0.5746069	−0.327287
crk	−0.3349415	0.0715481	−4.681	0.000	−0.4751733	−0.1947098
ant	0.3530984	0.0758826	4.653	0.000	0.2043712	0.5018256
col	0.368263	0.0833618	4.418	0.000	0.2048769	0.5316491
arm	0.4664409	0.0988147	4.720	0.000	0.2727676	0.6601141
ban	0.2097729	0.076176	2.754	0.006	0.0604708	0.359075
frm	0.6888737	0.0980013	7.029	0.000	0.4967947	0.8809528
_cut1	0.1488535	0.0834433		(Ancillary parameters)		
_cut2	2.04582	0.0869085				

y	Probability	Observed
1	Pr ($xb + u < \_cut1$)	0.4970
2	Pr ($\_cut1 < xb + u < \_cut2$)	0.3399
3	Pr ($\_cut2 < xb + u$)	0.1631

TABLE 2.14

Ordered Logit on Deprivation in Northern Ireland: Catholic Subsample

Ordered Logit Estimates Log Likelihood = −4283.363; Number of obs = 4364; $LR\chi^2(17)$ = 875.06; $Prob > \chi^2$ = 0.0000; $Pseudo\ R^2$ = 0.0927

y	Coefficient	Standard Error	z	P > |z|	[95% Conf. Interval]	
sex	−0.0430983	0.0621282	−0.694	0.488	−0.1648672	0.0786707
age	−0.0181642	0.0089546	−2.028	0.043	−0.0357148	−0.0006136
age2	0.00064	0.0001754	3.648	0.000	0.0002961	0.0009839
ret	0.6109521	0.1773859	3.444	0.001	0.2632821	0.958622
inac	1.487517	0.1299683	11.445	0.000	1.232784	1.742251
ue	1.364899	0.0906986	15.049	0.000	1.187133	1.542665
highed	−1.09291	0.1133639	−9.641	0.000	−1.315099	−0.8707208
mided	−0.7667852	0.2172056	−3.530	0.000	−1.1925	−0.3410701
hnum	0.9087931	0.0739094	12.296	0.000	0.7639333	1.053653
snpar	0.239649	0.0987062	2.428	0.015	0.0461884	0.4331096
ard	−0.5853662	0.16575	−3.532	0.000	−0.9102303	−0.2605022
crk	−0.3226664	0.1620424	−1.991	0.046	−0.6402637	−0.0050692
ant	0.2364382	0.1329838	1.778	0.075	−0.0242052	0.4970816
col	0.4795002	0.1039698	4.612	0.000	0.2757232	0.6832773
arm	0.2546794	0.0926459	2.749	0.006	0.0730968	0.4362621
ban	0.2025872	0.0954257	2.123	0.034	0.0155563	0.389618
frm	0.5243653	0.0950858	5.515	0.000	0.3380006	0.7107301
_cut1	0.0110645	0.1138485			(Ancillary parameters)	
_cut2	1.866844	0.1178967				

y	Probability		Observed			
1	Pr (xb + u < _cut1)		0.3824			
2	Pr (_cut1 < xb + u < _cut2)		0.3682			
3	Pr (_cut2 < xb + u)		0.2493			

39

contain only those variables with significantly non-zero coefficients? One argument is that if one believed a priori that a variable had a legitimate place in the equation specification then one should persist with this belief and include it, no matter what. Another argument, however, is that because the purpose of estimation and prediction is to confront equation specification with data, to base predictions on the coefficient estimates obtained from the full specification may be misleading since it would allow variables, whose legitimacy in the specification had been explicitly "rejected" by the data, to influence the predictions. In the previous section, following the first argument, the predictions were based on the full specification; in this section, following the second argument, they are based on a restricted specification.

Table 2.15, below, shows that: 50% and 38% respectively of the Protestant and Catholic samples were not deprived; 34% and 37% respectively of the Protestant and Catholic samples were mildly deprived; and 16% and 25% respectively of the Protestant and Catholic samples were severely deprived. The fact that a smaller proportion of Catholics, compared to Protestants, were not deprived and that a higher proportion of Catholics were both mildly and severely deprived (see Table 2.15) could be due to two reasons. First, those characteristics which increased the probability of being deprived were disproportionately concentrated among Catholics and/or those characteristics which decreased the probability of being deprived were disproportionately concentrated among Protestants.[32] Second, a particular attribute was penalized more harshly (if it was deprivation-increasing: for example, being unemployed) and/or was rewarded less generously (if it was deprivation-reducing: for example, having educational qualifications) if the person possessing the attribute was Catholic rather than Protestant.

TABLE 2.15
Deprivation in Northern Ireland by Religion

| | PERCENTAGE OF SAMPLE THAT IS: | | |
	Not Deprived	Mildly Deprived	Strongly Deprived
All persons	45.9	34.9	19.2
Catholics	38.2	36.8	24.9
Protestants	49.7	34.0	16.3

In order to determine how much of the Catholic-Protestant *deprivation gap* (defined as the difference in the proportions of Catholics and Protestants at different deprivation levels) was due to differences in characteristics, and how much was due to differences in coefficients, the econometric issue was posed in terms of the following questions:

1. What would have been the predicted probabilities of Protestants and Catholics being at different levels of deprivation if the characteristics possessed by each group had been evaluated using its own coefficient estimates? Denote these probabilities as, respectively, as $\hat{p}_{ij}^P$ and $\hat{p}_{ij}^C$ ($i = 1, \ldots, 13, 164$; $j = 1, 2, 3$) and their means as $\bar{p}_j^P$ and $\bar{p}_j^C$ ($j = 1, 2, 3$).[33]

2. What would have been the predicted probabilities of Protestants being at different levels of deprivation if their characteristics had been evaluated using *Catholic* coefficient estimates? Denote these "synthetic" probabilities as $\hat{q}_{ij}^P$ ($i = 1, \ldots, 13, 164$; $j = 1, 2, 3$) and their mean as $\bar{q}_j^P$ ($j = 1, 2, 3$).

3. How do the $\bar{q}_j^C$ compare to the $\bar{p}_j^C$ and $\bar{p}_j^P$?

The values of these three probabilities, $\bar{p}_j^P$, $\bar{q}_j^P$, and $\bar{p}_j^C$, are shown in Table 2.16, not just for Catholics and Protestants in their entirety but also for subgroups within Catholics and Protestants. Table 2.16 shows that, calculated over all the persons in the sample, the average "own-coefficient"[34] probabilities of Protestants being not deprived, mildly deprived and severely deprived were 50%, 34%, and 16% respectively. When Protestant characteristics were evaluated at Catholic coefficients, the probability of being not deprived *fell* to 46% and the probability of being severely deprived and of being mildly deprived *rose* to, respectively, 19% and 35%. This story was repeated when subgroups (single parents, retired persons, inactive persons, unemployed persons, and persons living in large families) from the sample were analyzed. With two exceptions (retired and unemployed persons) the probability of being not deprived always fell, and the probability of being severely deprived always rose, when Protestant characteristics were evaluated at Catholic coefficients though, naturally, the magnitude of these changes varied according to the subgroup being considered.[35] The largest fall (in the probability of being not deprived) and the sharpest rise (in the probability of being strongly deprived) was recorded for persons who were economically inactive. The pattern with respect to the probability of being mildly

TABLE 2.16
Predicted Probabilities of Catholics and Protestants Being at Different Deprivation Levels*

| | PREDICTED PROBABILITY OF BEING: | | |
	Not Deprived	Mildly Deprived	Strongly Deprived
All Persons			
Protestants at Protestant coefficients	49.7	34.0	16.2
Protestants at Catholic coefficients	45.8	35.3	18.9
Catholics at Catholic coefficients	38.2	36.9	24.9
Single Parents			
Protestants at Protestant coefficients	37.2	38.2	24.6
Protestants at Catholic coefficients	35.0	38.1	26.9
Catholics at Catholic coefficients	27.3	37.8	34.8
Retired			
Protestants at Protestant coefficients	24.8	41.4	33.8
Protestants at Catholic coefficients	24.6	40.7	• 34.6
Catholics at Catholic coefficients	20.7	39.6	39.7
Inactive			
Protestants at Protestant coefficients	31.3	41.4	27.3
Protestants at Catholic coefficients	20.7	38.9	40.5
Catholics at Catholic coefficients	16.5	37.6	45.9
Unemployed			
Protestants at Protestant coefficients	21.1	40.3	38.6
Protestants at Catholic coefficients	21.3	39.5	39.2
Catholics at Catholic coefficients	16.3	36.3	47.4
In Large Families			
Protestants at Protestant coefficients	28.1	40.0	31.9
Protestants at Catholic coefficients	26.0	39.0	34.9
Catholics at Catholic coefficients	23.2	38.6	38.2

*Calculated from the ordered logit estimates of Tables 2.13 and 2.14.

deprived was that the position of Protestant persons who were single parents, or retired, or unemployed or living in large families would have been unchanged,[36] but that of persons who were inactive would have worsened, had their characteristics been evaluated at Catholic coefficients.

The difference between the proportions of Protestants and Catholics in the different categories of deprivation, $\hat{p}_j^P - \hat{p}_j^C$, can be deconstructed as

$$\bar{p}_j^P - \bar{p}_j^C = (\bar{q}_j^P - \bar{p}_j^C) + (\bar{p}_j^P - \bar{q}_j^P) = A_j^* + B_j^* \qquad (2.20)$$

In Equation 2.20, A_j^* represents that part of the "deprivation gap" between Catholics and Protestants that is due to intergroup differences in characteristics, and B_j^* represents that part of the gap that is due to intergroup differences in coefficient values. These absolute differences can be expressed as proportions:

$$A_j = A_j^* / (\bar{p}_j^P - \bar{p}_j^C) \text{ and } B_j = 1 - A_j \qquad (2.21)$$

and the values of A_j and B_j are shown, in percentage form, in Table 2.17.

Table 2.17 shows that, computed over all the persons in the sample, 66% of the deprivation gap ($\bar{p}_j^P - \bar{p}_j^C$) between Catholics and Protestants, with respect to those who were not deprived, was due to the fact that persons in the Catholic subsample had characteristics which were different from persons in the Protestant subsample. Thirty four percent was due to the fact that Catholic characteristics were evaluated differently from Protestant characteristics. With respect to severe deprivation, 69% of the Catholic-Protestant gap was due to attribute differences, but with respect to mild deprivation only 55% of the Catholic-Protestant gap was due to attribute differences.

Turning to the specific subgroups, for persons who were single parents, retired, or unemployed, a comparatively large percentage of the Catholic-Protestant gap, with respect to the different levels of deprivation, was due to differences in characteristics between Catholics and Protestants. Relatively little of the gap was due to the fact that specific characteristics, when applied to Catholics, had more serious consequences for deprivation than they did when applied to Protestants.

TABLE 2.17

Contributions to the Deprivation Gap
Between Catholics and Protestants*

| | PERCENTAGE CONTRIBUTIONS BY DIFFERENCES IN: | |
	Characteristics (A_j)	Coefficients (B_j)
All Persons		
No-Deprivation	65.8	34.2
Mild-Deprivation	54.7	45.3
Strong-Deprivation	69.4	30.6
Single Parents		
No-Deprivation	78.2	21.8
Mild-Deprivation	64.4	35.6
Strong-Deprivation	77.7	22.3
Retired		
No-Deprivation	96.1	3.9
Mild-Deprivation	63.1	36.9
Strong-Deprivation	85.7	14.3
Inactive		
No-Deprivation	28.3	71.7
Mild-Deprivation	33.9	66.1
Strong-Deprivation	29.4	70.6
Unemployed		
No-Deprivation	103.0	−3.0
Mild-Deprivation	81.7	18.3
Strong-Deprivation	93.5	6.5
In Large Families		
No-Deprivation	58.1	41.9
Mild-Deprivation	27.2	72.8
Strong-Deprivation	51.4	48.6

*Calculated from the probability estimates of Table 2.16.

For example, for unemployed persons, over 94% of the Catholic-Protestant gap, with respect to severe deprivation, could be explained in terms of differences in characteristics between persons belonging to the two communities. However, for inactive persons differences in the coefficients used to evaluate characteristics (as opposed to differences in the characteristics themselves) accounted for over two-thirds of the deprivation gap between Catholics and Protestants. For persons living in large families, nearly half of the intercommunity deprivation gap in respect of severe deprivation and over 70% in respect of mild deprivation was due to differences in the coefficients used to evaluate the deprivation generating characteristics.

3. MULTINOMIAL LOGIT

Introduction

The previous chapter considered a class of models which were centered around events with multiple (>2) outcomes, where these outcomes were *inherently ordered*. In this class of models the dependent variable Y_i, when *defining these outcomes for person i* ($Y_i = 1$, for the first outcome; $Y_i = 2$, for the second outcome; and so on, until $Y_i = M$, for the Mth (last) outcome, $i = 1, \ldots, N$), was a discrete, ordinal variable. The appropriate methods of estimating such models were *ordered logit* and *ordered probit*. This chapter focuses on multiple outcome models where the outcomes are not ordered. The methodology of *multinomial logit*—which is the appropriate estimation method for this class of models—is explained and then applied to an analysis of occupation choice. An important property (and limitation) of multinomial logit is that of the *Independence of Irrelevant Alternatives (IIA)*. This chapter concludes with a discussion of this property and of how its limitations might be circumvented.

Several real world events provide examples of unordered outcomes. The choice of transportation to work (by bus, train or car) is clearly such an example. Other examples of unordered outcomes are occupational choice (unskilled; skilled; professional; managerial), choice of residence location (north; south; east; west), and choice of party at elections (Conservative; Liberal; Labor). A word that is common to all the above examples is "choice." Since there is nothing inherently good or bad about the outcomes, individuals may be viewed as choosing, from the menu of available outcomes, that outcome which suits them the best. Indeed, the framework of utility maximization offers a good starting point for understanding the structure of multiple outcome models when the outcomes are unordered.

A Random Utility Model

Given a choice between M alternatives (indexed, $j = 1, \ldots, M$), the utility that the ith person ($i = 1, \ldots, N$) derives from the jth alternative may be represented as U_{ij}. Suppose that this utility is a *linear* function of H factors (determining variables). Of these H factors, suppose that R factors are specific to the individual and have nothing to do with the nature of the choice and that S ($H = R + S$)

factors are specific to the choice and have nothing to do with the individual. For example, in terms of choosing a mode of travel to work, a characteristic of the individual might be that he or she does not live near a train station. The fact that this individual rarely travels to work by train has nothing to do with the nature of train travel but is solely an outcome of where that person lives. On the other hand, the fact that rush hour traffic is heavy is an attribute of car travel that reduces every person's utility from going to work by car. Suppose that the values of the R variables representing the *characteristics* of the ith person are $X_{ir}, r = 1 \dots, R$ and that the values of the variables representing the *attributes* of the jth choice are $W_{js}, s = 1, \dots, S$. The utility function may be written as

$$U_{ij} = \sum_{r=1}^{R} \beta_{jr} X_{ir} + \sum_{s=1}^{S} \gamma_{is} W_{js} + \varepsilon_{ij} = Z_{ij} + \varepsilon_{ij}, \qquad (3.1)$$

where β_{jr} is the coefficient associated with the rth characteristic ($r = 1, \dots, R$) for the jth alternative, γ_{is} is the coefficient associated with the sth attribute ($s = 1, \dots, S$) for the ith person, and

$$Z_{ij} = \sum_{r=1}^{R} \beta_{jr} X_{ir} + \sum_{s=1}^{S} \gamma_{is} W_{js}. \qquad (3.2)$$

An increase in X_{ir}, the value of the rth characteristic for person i, will cause his or her utility from choice j to rise if $\beta_{jr} > 0$ and to fall if $\beta_{jr} < 0$. An increase in W_{js}, the value of the sth attribute for choice j, will cause utility to rise for person i if $\gamma_{is} > 0$ and to fall if $\gamma_{is} < 0$. However, since the relationship between utility and its determining variables is not an exact one—for example, there may be factors left out of the equation or factors may be measured inaccurately—an error term, ε_i is included in the equation to capture this inexactitude. Hence the term random utility model.

A person will choose $j = m$ if and only if it offers, *of all the available choices*, the highest level of utility. In other words, if Y_i is a random variable whose value ($j = 1, \dots, M$) indicates the choice made by person i, the probability that person i will choose alternative m is

$$\Pr(Y_i = m) = \Pr(U_{im} > U_{ij}) \quad \text{for all } j = 1, \dots, M, \ j \neq m$$

$$\Longrightarrow \Pr(Z_{im} + \varepsilon_{im} > Z_{ij} + \varepsilon_{ij})$$

$$\Longrightarrow \Pr(\varepsilon_{ij} - \varepsilon_{im} < Z_{im} - Z_{ij}) \quad \text{for all } j = 1, \dots, M, \ j \neq m.$$

The Class of Logit Models: Multinomial and Conditional

McFadden (1973) has shown that if the M error terms ε_{ij} ($j = 1, \ldots, M$) are independently and identically distributed with Weibull distribution $F(\varepsilon_{ij}) = \exp[\exp(-\varepsilon_{ij})]$, then

$$\Pr(Y_i = m) = \frac{\exp(Z_{im})}{\sum_{j=1}^{M} \exp(Z_{ij})} \tag{3.3}$$

A model in which the probabilities of the different outcomes, $j = 1, \ldots, M$ are defined by Equation 3.3 is defined here as a *generalized logit* model. The term "generalized" conveys the fact that the model incorporates both characteristic effects and attribute effects, respectively the X_{ir} and W_{js} of Equation 3.1. Within the class of generalized logit models, two subclasses may be distinguished:

1. **Multinomial logit** models. These are models which incorporate only characteristic effects, so that for such models all the $\gamma_{is} = 0$ in Equation 3.1. In effect, such models apply when the data are individual specific.

2. **Conditional logit** models. These are models which incorporate only attribute effects, so that for such models all the $\beta_{jr} = 0$ in Equation 3.1. In effect, such models apply when the data are choice specific.

Multinomial Logit

The multinomial model is defined by Equation 3.3 but with the caveat that now, with the $\gamma_{is} = 0$,

$$Z_{ij} = \sum_{r=1}^{R} \beta_{jr} X_{ir}. \tag{3.4}$$

Because the probabilities $\Pr(Y_i = j)$ sum to 1 over all the choices (that is, $\sum_{j=1}^{M} \Pr(Y_i = j) = 1$), only $M - 1$ of the probabilities can be determined independently. Consequently the multinomial logit model of Equation 3.3 is indeterminate, as it is a system of M equations in only $M - 1$ independent unknowns. A convenient normalization that solves the problem is to set $\beta_{1r} = 0$, $r = 1, \ldots, R$. Under this normalization $Z_{i1} = 0$, and so from Equation 3.3

$$\Pr(Y_i = 1) = \frac{1}{1 + \sum_{j=2}^{M} \exp(Z_{ij})} \tag{3.5a}$$

$$\Pr(Y_i = m) = \frac{\exp(Z_{im})}{1 + \sum_{j=2}^{M} \exp(Z_{ij})} m = 2, \ldots, M. \tag{3.5b}$$

As a consequence of the normalization, the probabilities are uniquely determined so that Equation 3.5b represents a system of $M - 1$ equations in the $M - 1$ unknown probabilities, $\Pr(Y_i = m)$ with $\Pr(Y_i = 1)$ having being defined by Equation 3.5a through the normalization adopted.

From Equations 3.5a and 3.5b, the logarithm of the ratio of the probability of outcome $j = k$ to that of outcome $j = m$ is

$$\log\left(\frac{\Pr(Y_i = m)}{\Pr(Y_i = k)}\right) = \sum_{r=1}^{R} (\beta_{mr} - \beta_{kr}) X_{ir} = Z_{im} - Z_{ik}, \tag{3.6}$$

so that the logarithm of the *risk-ratio* (that is, the logarithm of the ratio of the probability of outcome m to that of outcome k, or $\log[\mathrm{Prob}(Y_i = m)/\mathrm{Prob}(Y_i = k)]$) does not depend upon the other choices. The risk-ratio or, as it is sometimes referred to, the *relative risk*—$\mathrm{Prob}(Y_i = m)/\mathrm{Prob}(Y_i = k)$—can easily be calculated from the log risk-ratio by taking its exponential. If $k = 1$, the log risk-ratio is

$$\log\left(\frac{\Pr(Y_i = m)}{\Pr(Y_i = 1)}\right) = \sum_{r=1}^{R} \beta_{mr} X_{ir} = Z_{im}, \quad (m = 2, \ldots, M), \tag{3.7}$$

and the risk-ratio is

$$\frac{\Pr(Y_i = m)}{\Pr(Y_i = 1)} = \exp\left(\sum_{r=1}^{R} \beta_{mr} X_{ir}\right)$$

$$= \exp(Z_{im}), \quad (m = 2, \ldots, M). \tag{3.8}$$

The risk-ratio (*RR*) should be distinguished from the *odds-ratio* (*OR*) where the latter refers to the probability of an outcome divided by 1—the probability of *that* outcome. Thus the odds-ratio for $j = m$ is

$$OR_m = \frac{\Pr(Y_i = m)}{1 - \Pr(Y_i = m)} = \frac{\Pr(Y_i = m)}{\Pr(Y_i = 1)} \frac{\Pr(Y_i = 1)}{1 - \Pr(Y_i = m)}$$

$$= \frac{RR_m \Pr(Y_i = 1)}{1 - RR_m \Pr(Y_i = 1)}, \tag{3.9}$$

where OR_m and RR_m are, respectively, the odds-ratio and the risk-ratio associated with outcome $j = m$ (the latter relative to the "base" outcome $j = 1$).

In a binary model, there is no distinction between the RR and the OR since the base outcome $Y_i = 1$ is simply the outcome $Y_i \neq m$. In a model with more than two possible outcomes, the outcomes $Y_i = 1$ and $Y_i \neq m$ are different. The natural method in all logit models is to express results as the ratio of the likelihood of an outcome and the likelihood of some base outcome—that is, to compute the RR. In binary models, however, the RR is the OR and so results in such models are expressed in terms of the latter. In multinomial logit models, on the other hand, results are expressed in terms of RR and not in terms of OR since these are now different from each other.

Estimation and Prediction

Each of the N observations on the dependent variable Y_i ($i = 1, \ldots, N$) is treated as a single draw from a multinomial distribution with M outcomes. Define a dummy variable $\delta_{ij} = 1$ if person i makes choice j, $\delta_{ij} = 0$ otherwise, $j = 1, \ldots, M$. Then the likelihood of observing the sample is

$$L = \prod_{i=1}^{N} \prod_{j=1}^{M} [\Pr(Y_i = j)]^{\delta_{ij}} \Rightarrow \log L = \sum_{i=1}^{N} \sum_{j=1}^{M} \delta_{ij} \Pr(Y_i = j), \tag{3.10}$$

where $\Pr(Y_i = j)$ is defined by Equation 3.5a if $j = 1$, and by Equation 3.5b if $j > 1$. The parameter estimates $\hat{\beta}_{jr}$ ($j = 1, \ldots, M$; $r = 1, \ldots, R$) are chosen so as to maximize the likelihood function (Equation 3.10).

Given these estimates, for each person $i = 1, \ldots, N$ one can form estimates of Z_{ij} using Equation 3.4, with the $\hat{\beta}_{jr}$ in place of β_{jr}, for every outcome $j = 1, \ldots, M$. Then, using these estimates, $\widehat{Z}_{ij}$, the predicted probabilities, $\hat{p}_{ij}$, can be computed from Equations 3.5a and 3.5b, for $i = 1, \ldots, N$ and $j = 1, \ldots, M$. Note that for every person, the estimated probabilities must sum to unity across all the outcomes, $\sum_{j=1}^{M} \hat{p}_{ij} = 1$. A property of the multinomial logit model is that the mean of the estimated individual probabilities for each outcome ($\bar{p}_j = \sum_{i=1}^{N} \hat{p}_{ij}, j = 1, \ldots, M$) is equal to the observed proportion of persons in that outcome category.

An alternative way of predicting probabilities from the multinomial logit model is to compute the mean of Z_{ij} across all the persons for each outcome $j = 1, \ldots, M$ as

$$\bar{Z}_j = \sum_{i=1}^{N} Z_{ij}/N = \sum_{i=1}^{N}\left(\sum_{r=1}^{R} \beta_{jr}X_{ir}\right)\Big/N$$

$$= \sum_{r=1}^{R} \beta_{jr}\left(\sum_{i=1}^{N} X_{ir}/N\right) = \sum_{r=1}^{R} \beta_{jr}\bar{X}_r, \tag{3.11}$$

and then to calculate the predicted probabilities as

$$\hat{p}_1 = \frac{1}{1 + \sum_{j=2}^{M} \exp(\bar{Z}_j)} \tag{3.12a}$$

$$\hat{p}_m = \frac{\exp(\bar{Z}_j)}{1 + \sum_{j=2}^{M} \exp(\bar{Z}_j)} \quad m = 2, \ldots, M. \tag{3.12b}$$

Then, in general, $\bar{p}_1 \neq \hat{p}_1$, $\bar{p}_2 \neq \hat{p}_2$, and $\bar{p}_3 \neq \hat{p}_3$, which is not surprising because although both the $\bar{p}_j$ and the $\hat{p}_j$ ($j = 1, 2, 3$) purport to measure the overall probability of the different outcomes, they are computed very differently. The $\bar{p}_j$ are computed as the mean of the estimated individual probabilities, $\hat{p}_{ij}$, and the $\hat{p}_{ij}$ are computed from the values of the determining variables for the individual. On the other hand the $\hat{p}_j$, which bypass the individual probabilities, are computed from the mean values $\bar{Z}_j$ of the individual Z_{ij} (or, equivalently, calculated using the mean values $\bar{X}_r$ of the determining variables X_{ir}).

Marginal Effects

The question of the marginal effect on the probabilities of the different outcomes of a small change in the value of the determining variables can be phrased in two separate ways. For a small change in

X_{ir} (the value, for person i, of the rth determining variable) what will be the change, for some outcome m, in the following?

1. The probability $\Pr(Y_i = m)$.
2. The risk-ratio $\Pr(Y_i = m)/\Pr(Y_i = 1)$.

The second question is relatively easy to answer, but the first is much more difficult. Taking the easier question first, from Equation 3.7,

$$\frac{\partial}{\partial X_{ir}} \log\left(\frac{\mathrm{Prob}(Y_i = m)}{\mathrm{Prob}(Y_i = 1)}\right) = \beta_{mr},$$

so that for a small change in X_{ir} the direction of change in the risk-ratio can be inferred from the sign of the associated coefficient; the *relative* probability of $j = m$ increases if $\beta_{mr} > 0$ and decreases if $\beta_{mr} < 0$.

However, the direction of change in $\Pr(Y_i = m)$, the probability of observing outcome $j = m$, for a small change in X_{ir} cannot be inferred from the sign of β_{mr}. The reason is that in a multinomial model a change in the value of a variable for a particular person affects for him or her the probability of *every* outcome. Since these probabilities are constrained to sum to unity, whether $\Pr(Y_i = m)$ goes up or down depends upon what happens to the other probabilities. Therefore in effect it depends not just upon the sign of β_{mr} but also upon the size of that coefficient relative to the size of the other coefficients attached to the variable, that is, to the $\beta_{jr} j = 1, \ldots, M, j \neq m$. Consequently, $\partial \mathrm{Prob}(Y_i = m)/\partial X_{ir}$ need not have the same sign as β_{mr}.

The most effective way of establishing the effect of a change in the value of a variable upon the outcome probabilities in a multinomial model is to compare the computed probabilities before and after the change *with the values of the other variables left unchanged*. This method is most useful when the determining variable under analysis is a dummy variable. If the rth variable is a dummy variable, so that $X_{ir} = 1$ or $X_{ir} = 0$, then first evaluate Z_{ij} for every outcome $j = 1, \ldots, M$ using Equation 3.4, under the assumption that for all persons $X_{ir} = 1$. Call this estimate $\widehat{Z}_{ij}^1$. Then ceteris paribus evaluate Z_{ij} for every outcome $j = 1, \ldots, M$, under the assumption that for all persons $X_{ir} = 0$. Call this estimate $\widehat{Z}_{ij}^0$ and note that $\widehat{Z}_{ij}^1 = \widehat{Z}_{ij}^0 + \hat{\beta}_{jr}$.

Using Equations 3.5a and 3.5b to compute the predicted probabilities, first using $Z_{ij} = \widehat{Z}^1_{ij}$ (to obtain $\hat{p}^1_{ij}$) and then using $Z_{ij} = \widehat{Z}^0_{ij}$ (to obtain $\hat{p}^0_{ij}$). Denote the mean probabilities, computed over all persons, as $\bar{p}^1_j = \sum_{i=1}^{N} \hat{p}^1_{ij}$ and $\bar{p}^0_j = \sum_{i=1}^{N} \hat{p}^0_{ij}$. The difference between the $\bar{p}^1_j$ and $\bar{p}^0_j$ is the "mean" effect of a change, for all persons, in the value of the rth determining variable on the probability of observing outcome j ($j = 1, \ldots, M$).

An alternative way of keeping the values of the other variables unchanged while changes to the value of a variable are being analyzed is to set the values of the other variables to their mean values, $\bar{X}_r = \sum_{i=1}^{N} X_{ir}/N$. Then define

$$\bar{Z}^1_j = \beta_{jr} + \sum_{s \neq r} \beta_{js} \bar{X}_s \text{ and } \bar{Z}^0_j = \sum_{s \neq r} \beta_{js} \bar{X}_s, \quad (3.13)$$

and use Equations 3.5a and 3.5b to compute the predicted probabilities, first using $Z_{ij} = \bar{Z}^1_j$ (to obtain $\hat{p}^1_j$) and then using $Z_{ij} = \bar{Z}^0_j$ (to obtain $\hat{p}^0_j$). The difference between the $\hat{p}^1_j$ and $\hat{p}^0_j$ is the effect of a change in the value of the rth determining variable for the "average person" (defined as a person with average values for all other variables) on the probability of observing outcome j ($j = 1, \ldots, M$) for that person.

The effect of a change in the value of a determining variable upon the risk-ratio is much more straightforward. The difference in the log risk-ratio when, respectively, $X_{ir} = 1$ and $X_{ir} = 0$ is, from Equation 3.7, $\widehat{Z}^1_{ij} - \widehat{Z}^0_{ij} = \hat{\beta}_{jr}$, and so the corresponding difference in the risk-ratio is $\exp(\hat{\beta}_{jr})$. Thus the exponential value of a coefficient represents the change in the risk-ratio (for that outcome) for a *one unit* change in a determining variable. Needless to say, a one unit change is most appropriately considered for variables that are dummy variables.

Application to Occupational Outcomes

Equation Specification

Information was extracted from the 1991 Census for Britain on the occupational class (and other characteristics) of male full-time employees who were between the ages of 25 and 45 years.[37] The occupational classes were unskilled/semiskilled (*UNS*), skilled manual/nonmanual (*SKL*), and professional/managerial/technical (*PMT*). Of these

TABLE 3.1

Sample Statistics of Male Full-Time Employees, by Ethnic Group*

	White	Black	Indian
Sample Size	96,297	863	1,572
Age (yrs)	35.4	32.6	34.9
% in Occupational Class			
Professional/Managerial/Technical	41.7	29.3	35.1
Skilled Manual/Nonmanual	40.7	50.2	39.8
Unskilled/Semiskilled	17.6	20.5	25.2
% With Post-18 Qualification			
Degree	15.1	7.9	17.4
Subdegree	9.2	7.1	6.6
No Post-18 Qualification	75.7	85.0	76.0
% Born in			
Britain	96.2	54.6	7.6
Overseas	3.8	45.4	92.4
% Living in			
London	10.1	51.5	43.6
North	50.3	24.9	30.2
South	39.6	23.6	26.2

*Information from the 1991 census for Britain.

98,732 men 96,297 were "white," 863 were "black," and 1,572 were (Asian) Indians. Table 3.1 shows the salient features of the sample statistics. This table demonstrates very clearly the differences in occupational status between whites, blacks, and Indians. For example, while 42% of white male employees were in PMT jobs, only 35% of Indians, and only 29% of blacks were in similar employment. Nor could these differences be explained away in terms of differences in characteristics: 17% of Indian employees had degrees, as opposed to 15% of white employes.

In terms of the choice formulation set out earlier, it would appear that whites, blacks, and Indians face different sets of constraints in making their choices. But such observations beg the question of what lies behind these differences in constraints. It is possible that persons from minority ethnic groups face disadvantage in the labour market in relation to equally qualified white persons. It is also possible that persons from minority ethnic groups have less favorable worker characteristics (hereafter, referred to as characteristics) than white persons. Intergroup differences in occupational representation may, therefore, be the result of both *ethnic* and *characteristics* disadvantage. The crucial question is: how much of these differences is the result of ethnic disadvantage and how much is due to characteristics disadvantage? This section shows how the method of multinomial logit that is described in the previous section can be used to answer this question.

The starting point was to define the dependent variable Y_i for each of the $i = 1, \ldots, 98,732$ men in the sample such that:

- $Y_i = 1$ if the person was employed in a UNS occupation.
- $Y_i = 2$ if the person was employed in a SKL occupation.
- $Y_i = 3$ if the person was employed in a PMT occupation.

The determining variables used in the multinomial specification were:

Age

- AGE in years: normalized by setting $AGE = 0$ for persons who were 25 years old.

Education[38]

- HED $= 1$, if the person had degree-level qualifications; HED $= 0$, otherwise.
- MED $= 1$, if the person had post-A level, but less than degree, qualifications; MED $= 0$, otherwise.

Area of Study[39]

- SCI $= 1$, if area was science related; SCI $= 0$, otherwise.
- BUS $= 1$, if area was business-studies related; BUS $= 0$, otherwise.

Ethnicity[40]

- BLK $= 1$, if the person's ethnicity was Black-Caribbean; BLK $= 0$, otherwise.
- IND $= 1$, if the person's ethnicity was Indian; IND $= 0$, otherwise.

Country of Birth

- OVB $= 1$, if the person was born outside Britain; OVB $= 0$, otherwise.

Area of Residence[41]

- STH = 1, if living in the South of Britain[42] (excluding London); STH = 0, otherwise.
- NTH = 1, if living in the North of Britain;[43] NTH = 0, otherwise.

On the basis of these variables the two (log) risk-ratio equations (Equation 3.7 of the previous section, for $j = 2, 3$) were specified as follows:

$$\log\left(\frac{\Pr(Y_i = j)}{\Pr(Y_i = 1)}\right) = \alpha_{j0} + \beta_{j0} * BLK_i + \gamma_{j0} * IND_i \qquad (3.14)$$

$$+ \alpha_{j1} * OVB_i + \beta_{j1} * BLK_i * OVB_i + \gamma_{j1} * IND_i * OVB_i$$

$$+ \alpha_{j2} * HED_i + \beta_{j2} * BLK_i * HED_i + \gamma_{j2} * IND_i * HED_i$$

$$+ \alpha_{j3} * MED_i + \beta_{j3} * BLK_i * MED_i + \gamma_{j3} * IND_i * MED_i$$

$$+ \alpha_{j4} * NTH_i + \beta_{j4} * BLK_i * NTH_i + \gamma_{j4} * IND_i * NTH_i$$

$$+ \alpha_{j5} * STH_i + \beta_{j5} * BLK_i * STH_i + \gamma_{j5} * IND_i * STH_i$$

$$+ \theta_{j1} * AGE_i + \theta_{j2} * AGE_i^2 + \theta_{j3} * BUS_i + \theta_{j4} * SCI_i$$

$$+ \theta_{j5} * BUS_i * HED_i + \theta_{j6} * SCI_i * HED_i = Z_{ij}$$

The outcome $j = 1$—that is, being in the UNS occupational class—is hereafter referred to as the "base" outcome. The coefficients of this outcome are set to zero and the risk-ratios of the other outcomes are defined with respect to the probability of this base outcome.[44]

In Equation 3.14, the α_{jr} are the "white" coefficients; the β_{jr} and the γ_{jr} represent the *additional* contribution to these coefficients resulting from being, respectively, black and Indian. On the other hand, the θ_{jk} are coefficients that are assumed not have an "ethnic dimension" meaning that they are (assumed to be) invariant with respect to ethnicity. From Equation 3.14, the log risk-ratio is α_{j0} for a 25-year-old (AGE = 0), British-born (OVB = 0), white (BLK = IND = 0) male employee with no post-18 educational qualifications (HED = MED = 0), living in London (NTH = STH = 0). If such a person were, instead, black or Indian then his log risk-ratio would *change* by, respectively β_{j0} and γ_{j0} to become, respectively: $(\alpha_{j0} + \beta_{j0})$ and $(\alpha_{j0} + \gamma_{j0})$.

If, for example, $\beta_{j0} < 0$ then the ratio of the probability of being in occupational class j to the probability of being in the UNS class would be higher for a white person than for an equivalent black person. The magnitude of the coefficients β_{j0} and γ_{j0} measure the degree of "ethnic disadvantage" (with respect to occupational class j) faced by blacks and Indians vis-à-vis whites in the context of the characteristics set out above.[45] The interaction terms in Equation 3.14, involving the ethnic variables *BLK* and *IND*, allow the degree of ethnic disadvantage to vary with some of the nonethnic characteristics of a person. These nonethnic characteristics were country of birth, region of residence, and level of qualification. For men with the characteristics described above, the log risk-ratio of a white overseas born (OVB = 1) male being in occupational class j is $\alpha_{j0} + \alpha_{j1}$, and that of a black overseas born (OVB = 1) male is $\alpha_{j0} + \alpha_{j1} + \beta_{j0} + \beta_{j1}$.

A set of nonethnic interactions that were included in Equation 3.14 was between the area of study and the level of qualification. These interactions have associated coefficients θ_{j5} and θ_{j6} and nonzero values, for these would imply that in determining the value of the risk-ratio it was not just the area of study and the level of qualification, considered separately, that mattered but also how subject and qualifications fused to produce, for example, a graduate with a science-related degree.

The Equation Statistics

Equation 3.14 was estimated first without any restrictions imposed upon its coefficients and then with some of its coefficients constrained to be zero. Equation 3.14 contained a total of 48 coefficients, 24 each in the SKL ($j = 1$) and PMT ($j = 2$) equations. Of these 48 coefficients, 13 coefficients were set to zero. These coefficients set to zero were, individually, not significantly different from zero, and likelihood ratio tests with $\chi^2(13) = 8.1$ did not reject the joint hypotheses that they were all equal to zero. Tables 3.2 and 3.3 show, respectively, the results of estimating Equation 3.14 without and with the zero restrictions imposed. A comparison of the results with and without the zero restrictions imposed showed that imposing the zero restrictions did not qualitatively affect the estimates of the coefficients that were *not* set to zero. The z-ratios in Tables 3.2 and 3.3 are the ratios of the estimated coefficients to their estimated standard errors the z-ratios are asymptotically distributed as $N(0, 1)$ under the null hypothesis that the associated coefficients are zero.[46]

Greene (2000, p. 831–833) has a number of suggestions for measuring the "goodness-of-fit" of equations with discrete dependent variables. At a minimum he suggests that one should report the maximized value of the log-likelihood function. The values of L_1 are the maximized log-likelihood values shown at the head of the tables (respectively, −86948.078 and −86952.126). Since the hypothesis that all the slopes in the model are zero is often interesting, the results of comparing the "full" model with an "intercept only" model should also be reported. The χ^2 values at the head of the Tables 3.2 and 3.3 (30880.36 and 30872.27) are defined as $2(L_1 - L_0)$ where L_0 is the value of the log-likelihood function when the only explanatory variable was the constant term and L_1, as observed earlier is the value of the log-likelihood function when all the explanatory variables were included. The degrees of freedom are equal the number of slope coefficients estimated. These χ^2 values decisively reject the null hypothesis that the model did not have greater explanatory power than an "intercept only" model.

The "pseudo-R^2" is defined as $1 - L_1/L_0$ and is due to McFadden (1973). This is bounded from below by 0 and from above by 1. A 0 value corresponds to all the slope coefficients being zero, and a value of 1 corresponds to perfect prediction (that is, to $L_1 = 0$). Unfortunately, as Greene (2000) notes, the values between 0 and 1 have no natural interpretation, though it has been suggested that the pseudo-R^2 value increases as the fit of the model improves. Other measures have been suggested. Ben-Akiva and Lerman (1985) and Kay and Little (1986) suggested a fit measure that measured the average probability of correct prediction by the prediction rule. Cramer (1999) suggested a measure that corrected for the failure of Ben-Akiva/Lerman measure to take into account that, in unbalanced samples, the less frequent outcome will usually be predicted very badly. In their survey of pseudo-R^2 measures, Veall and Zimmerman (1996) argued that in models of the multinomial probit or logit type, only the McFadden (1973) measure "seemed worthwhile."

An alternative to "point" measures of goodness-of-fit might be to assess the predictive ability of the model. Such assessments are routine in models of binary choice where the hits ($Y_i = 1$) and misses ($Y_i = 0$) predicted by the model *on the basis of a prediction rule* (e.g., $Y_i = 1$ if $\hat{p}_i > 0.5$, $Y_i = 0$, otherwise) are compared to the actual hits and misses. This procedure could be extended to

(text continues on page 62)

TABLE 3.2

Multinimial Logit Estimation of Occupational Choice: Full Specification

Multinominal Regression Log Likelihood $= -86948.078$; Number of obs $= 98732$; LR $\chi^2(46) = 30880.36$; Prob $> \chi^2 = 0.0000$; Pseudo $R^2 = 0.1508$

y	Coefficient	Standard Error	z	P > |z|	[95% Conf. Interval]	
y = 2						
age	0.0261899	0.0064884	4.036	0.000	0.0134729	0.0389069
age2	-0.0009772	0.000306	-3.194	0.001	-0.0015768	-0.0003775
blkcb	0.3655553	0.1672964	2.185	0.029	0.0376603	0.6934503
blkovs	0.08195	0.1887671	0.434	0.664	-0.2880266	0.4519266
ind	0.4707812	0.2790797	1.687	0.092	-0.0762051	1.017767
indovs	-0.6064709	0.2728072	-2.223	0.026	-1.141163	-0.0717786
ovsbn	-0.109104	0.0540097	-2.020	0.043	-0.214961	-0.003247
north	-0.1445068	0.0355612	-4.064	0.000	-0.2142055	-0.0748082
indnth	-0.4185296	0.1586727	-2.638	0.008	-0.7295225	-0.1075368
blknth	-0.7146398	0.2170033	-3.293	0.001	-1.139959	-0.2893211
south	-0.0776431	0.0365635	-2.124	0.034	-0.1493062	-0.0059799
indsth	-0.3459856	0.1654501	-2.091	0.037	-0.6702619	-0.0217094
blksth	-0.6664942	0.225652	-2.954	0.003	-1.108764	-0.2242244
highed	1.083631	0.1822909	5.945	0.000	0.7263477	1.440915
indhe	-0.2140724	0.4490211	-0.477	0.634	-1.094138	0.6659928
blkhe	-1.26384	0.7230326	-1.748	0.080	-2.680958	0.1532775
mided	0.6060102	0.2709862	2.236	0.025	0.0748871	1.137133
indme	-0.3332887	0.4428193	-0.753	0.452	-1.201199	0.5346213
blkme	0.9548953	1.034856	0.923	0.356	-1.073386	2.983177
subbus	1.205355	0.3189165	3.780	0.000	0.5802901	1.83042
subsci	0.5103959	0.2820694	1.809	0.070	-0.0424499	1.063242
bush	-0.6332213	0.4138681	-1.530	0.126	-1.444388	0.1779454
scih	-0.795496	0.3545109	-2.244	0.025	-1.490325	-0.1006674
_cons	0.7755227	0.042255	18.353	0.000	0.6927043	0.858341

y = 3

	Coef.	Std. Err.	z	P>|z|	[95% Conf.	Interval]
age	0.0893431	0.0073103	12.222	0.000	0.0750152	0.103671
age2	-0.0024609	0.0003397	-7.244	0.000	-0.0031266	-0.0017951
blkcb	-0.0672642	0.1950314	-0.345	0.730	-0.4495187	0.3149904
blkovs	-0.5708287	0.2236755	-2.552	0.011	-1.009225	-0.1324329
ind	0.0918407	0.3348558	0.274	0.784	-0.5644646	0.748146
indovs	-0.7893394	0.3292557	-2.397	0.017	-1.434669	-0.1440101
ovsbn	0.0932756	0.0565007	1.651	0.099	-0.0174637	0.204015
north	-0.6113055	0.037637	-16.242	0.000	-0.6850727	-0.5375384
indnth	-0.4496822	0.1943706	-2.416	0.016	-0.8506416	-0.0887228
blknth	-0.1352359	0.2598729	-0.520	0.603	-0.6445774	0.3741057
south	-0.1697707	0.0383309	-4.429	0.000	-0.244898	-0.0946434
indsth	-0.5134623	0.1951871	-2.631	0.009	-0.8960221	-0.1309025
blksth	-0.418329	0.268549	-1.558	0.119	-0.9446753	0.1080174
highed	3.877128	0.1713991	22.620	0.000	3.541191	4.213064
indhe	-0.1698475	0.4062421	-0.418	0.676	-0.9660674	0.6263724
blkhe	-1.139974	0.6128429	-1.860	0.063	-2.341124	0.0611758
mided	2.850769	0.2490918	11.445	0.000	2.362558	3.33898
indme	-0.2049551	0.4154452	-0.493	0.622	-1.019213	0.6093026
blkme	0.743211	1.023254	0.726	0.468	-1.262329	2.748751
subbus	0.8151578	0.298235	2.733	0.006	0.230628	1.399688
subsci	0.2500103	0.2599529	0.962	0.336	-0.2594879	0.7595085
bush	-0.0414593	0.3898265	-0.106	0.915	-0.8055052	0.7225865
scih	0.1600076	0.3281777	0.488	0.626	-0.4832088	0.803224
_cons	-0.0164003	0.0462775	-0.354	0.723	-0.1071026	0.0743019

(Outcome y = 1 is the comparison group).

59

TABLE 3.3

Multinimial Logit Estimation of Occupational Choice: Restricted Specification

Multinomial Regression Log Likelihood = −86952.126; Number of obs = 98732; LR $\chi^2(33)$ = 30872.27; Prob > χ^2 = 0.0000; Pseudo R^2 = 0.1508

y	Coefficient	Standard Error	z	P > |z|	[95% Conf. Interval]	
y = 2						
age	0.026323	0.0064795	4.062	0.000	0.0136234	0.0390227
age2	-0.0009813	0.0003057	-3.210	0.001	-0.0015805	-0.0003822
blk	0.426683	0.1116093	3.823	0.000	0.2079328	0.6454332
blkovs	(dropped)					
ind	0.4144723	0.2152989	1.925	0.054	-0.0075059	0.8364504
indovs	-0.565894	0.2215624	-2.554	0.011	-1.000148	-0.1316396
ovsbn	-0.106519	0.05151	-2.068	0.039	-0.2074767	-0.0055613
north	-0.1438706	0.0351996	-4.087	0.000	-0.2128606	-0.0748806
indnth	-0.4152823	0.1546391	-2.685	0.007	-0.7183694	-0.1121952
blknth	-0.6568392	0.1761849	-3.728	0.000	-1.002155	-0.3115232
south	-0.0765746	0.0362333	-2.113	0.035	-0.1475906	-0.0055585
indsth	-0.3417737	0.1632409	-2.094	0.036	-0.66172	-0.0218275
blksth	-0.6677784	0.2086548	-3.200	0.001	-1.076734	-0.2588226
highed	1.373433	0.1088435	12.618	0.000	1.160104	1.586763
indhe	(dropped)					
blkhe	-1.312484	0.7222905	-1.817	0.069	-2.728147	0.1031797
mided	0.8347366	0.1476385	5.654	0.000	0.5453704	1.124103
indme	(dropped)					
blkme	(dropped)					
subbus	0.9348005	0.1923779	4.859	0.000	0.5577467	1.311854
subsci	0.2707989	0.1355025	1.998	0.046	0.0052189	0.5363788
bush	-0.6003602	0.1675312	-3.584	0.000	-0.9287153	-0.272005
scih	-0.9608239	0.1601125	-6.001	0.000	-1.274639	-0.6470093
_cons	0.7737729	0.041826	18.500	0.000	0.6917954	0.8557504

$y = 3$

	Coef.	Std. Err.	z	P>|z|	[95% Conf.	Interval]
age	0.0895118	0.0073026	12.258	0.000	0.0751989	0.1038247
age2	-0.0024658	0.0003395	-7.263	0.000	-0.0031313	-0.0018004
blk	(dropped)					
blkovs	-0.6781675	0.154667	-4.385	0.000	-0.9813092	-0.3750257
ind	(dropped)					
indovs	-0.7063556	0.1295174	-5.454	0.000	-0.9602051	-0.4525062
ovsbn	0.0954136	0.0553704	1.723	0.085	-0.0131104	0.2039377
north	-0.6105677	0.0368792	-16.556	0.000	-0.6828497	-0.5382858
indnth	-0.4542501	0.181704	-2.500	0.012	-0.8103834	-0.0981168
blknth	(dropped)					
south	-0.1682364	0.0376312	-4.471	0.000	-0.2419921	-0.0944807
indsth	-0.501347	0.1887684	-2.656	0.008	-0.8713262	-0.1313678
blksth	-0.4365967	0.2228379	-1.959	0.050	-0.8733511	0.0001576
highed	4.171942	0.0844821	49.383	0.000	4.00636	4.337524
indhe	(dropped)					
blkhe	-1.198389	0.6072452	-1.973	0.048	-2.388567	-0.0082099
mided	3.089609	0.0695418	44.428	0.000	2.95331	3.225909
indme	(dropped)					
blkme	(dropped)					
subbus	0.5326858	0.1343444	3.965	0.000	0.2693757	0.795996
subsci	(dropped)					
bush	(dropped)					
scih	(dropped)					
_cons	-0.0188423	0.0454545	-0.415	0.678	-0.1079314	0.0702469

(Outcome $y = 1$ is the comparison group).
Mlogit: likelihood-ratio test $\chi^2(13) = 8.10$; Prob $> \chi^2 = 0.8372$.

61

multiple outcome models, where the predictions could be based on a rule whereby $Y_i = m$, if $\hat{p}_{im} = \text{Max}_j(\hat{p}_{ij})$. These predictions could then be compared to a "naive" model that predicted all cases to be in the modal category of the dependent variable, and the percentage reduction in error in moving from the naive to the full model could be computed. This approach is, however, not without pitfalls. First, unlike the case of the linear regression model, where the coefficients are chosen to maximize R^2, in discrete choice models the coefficient estimates do not maximize *any* goodness-of-fit measure. Therefore to assess the model on the basis of goodness-of-fit, however measured, may be misleading. Second, the predictions are critically dependent on the prediction rule adopted, and the adopted rule may turn out to be quite inappropriate to the needs of the sample. For example, if the sample is unbalanced in a binary model—that is, has many more 1s than 0s—then a rule that the model should predict the outcome for which the estimated probability is greatest might never predict a 1 (or a 0).

The Estimates

As the earlier discussion emphasized, the sign of a coefficient estimate in Tables 3.2 and 3.3 reflects the direction of change in the risk-ratio, $\Pr(Y_i = j)/\Pr(Y_i = 1)$, in response to a ceteris paribus change in the value of the variable to which the coefficient is attached. It does not reflect the direction of change in the individual probabilities $\Pr(Y_i = j)$. The estimation results reported in this subsection pertain to the *restricted* coefficients shown in Table 3.3. This was because when the full equation specification (as shown in Table 3.2) was confronted with the data, it was found that only a subset of these variables exerted a significant effect on the risk-ratios. In most cases, the variables that were excluded were interaction terms involving a nonethnic variable, X, and the ethnic variables, *BLK* and/or *IND*. This meant that while the variable X by itself exerted a significant effect on the log risk-ratio of person i belonging to a particular occupational class, the ethnicity of a person did not alter this effect.

The estimation results, shown in Table 3.3, identify three characteristics as being important for improving a person's risk-ratio of being in SKL or PMT employment:[47]

- living in London;
- having post-18 educational qualifications, preferably a degree and preferably in a business-related subject; and
- being born in Britain.

Living in London conferred two benefits. First, there was a general benefit that accrued to *all* persons. This stemmed from the fact that the risk-ratio of being in SKL or in PMT employment of men living in the North or in the South was ceteris paribus lower than that of persons living in the London[48] ($\hat{\alpha}_{j4}, \hat{\alpha}_{j5} < 0$, $j = 2, 3$ in Equation 3.14). Second, there was a specific benefit that accrued to blacks and Indians. By living in London, rather than outside London, blacks and Indians experienced a greater boost to their risk-ratio of SKL and PMT employment than did whites. The reduction in the risk-ratio, of being in SKL or PMT employment, for a person living outside London, relative to living in London, would have been greater if that person had been black[49] or Indian than if that person had been white($\hat{\beta}_{j4}, \hat{\beta}_{j5}, \hat{\gamma}_{j4}, \hat{\gamma}_{j5} < 0$, $j = 2$ and $\hat{\beta}_{j5}, \hat{\gamma}_{j4}, \hat{\gamma}_{j5} < 0$, $j = 3$).

In that sense, London was more kind to Indians and blacks, relative to whites, than was the rest of Britain. In particular, Indians living in London ($NTH_i = STH_i = 0$) did not face any disadvantage relative to whites, with respect to the risk-ratio of being in PMT employment since $\hat{\gamma}_{j0} = 0$, $j = 3$. Living outside London (e.g., $NTH_i = 1$) reduced this risk-ratio for all persons, but the reduction was greater for Indians than for whites. In that sense, the ethnic parity in PMT employment that Indians enjoyed with respect to whites in London was eroded outside London. The "London effect" operated most strongly in favor of blacks. Only 10% of white male full-time employees lived in London, but it was the area of residence of 44% of Indian, and 53% of black, male full-time employees.

Being born outside Britain was always a disadvantage: the risk-ratio of being in SKL and in PMT employment was greater for British-born than for overseas-born men ($\hat{\alpha}_{j1} < 0$, $j = 2, 3$ in Equation 3.14). However, given that the interaction of the birthplace variable with ethnicity was negative ($\hat{\beta}_{j1}, \hat{\gamma}_{j1} < 0$, $j = 2, 3$), the disadvantage of being born overseas was greater for Indians and blacks than it was for whites, i.e., the reduction in a person's risk-ratio that stemmed from being born overseas was greater for Indians and blacks than for whites.

Post-18 qualifications, both in the form of sub-degree and of degree-level (or higher) qualifications, raised the risk-ratio of being in a

SKL or PMT job, with the effect of degree-level qualifications being stronger than that of sub-degree qualifications. In addition to qualifications, the subject in which the qualification was obtained also mattered. Qualifications in business-studies type subjects provided the best means of entry into both SKL and PMT employment. The econometric finding was that for Indian men, γ_{j2} (HED) and γ_{j3} (MED) in Equation 3.14 were not significantly different from zero for both SKL ($j = 1$) and PMT ($j = 2$) employment. For black men, β_{j2} (HED) was significantly negative for PMT employment. Black men obtained a lower return on their degree-level qualifications than did Indian or white men, but Indian and white men received the same return.

The literature suggests that immigrants, particularly from developing countries, who had obtained their qualifications in their countries of origin suffered (whether justifiably or not) from a perception that such qualifications were "less worthy" than equivalent qualifications obtained in the host country. Unfortunately the data do not record, in the case of persons born overseas, their date of arrival in Britain. Consequently, while it is known that the majority of the black and Indian men in the sample were born overseas (Table 3.1) there is no information on their age at arrival in Britain and, by implication, no possibility of surmising where they might have obtained their post-18 qualifications (if any).

The Predicted Probabilities

Using the estimated $\widehat{Z}_{ij}$—which, remembering that $\widehat{Z}_{ij} = \sum_{r=1}^{R} \hat{\beta}_{jr} \times X_{ir}$, are computed using the estimates shown in Table 3.3 in conjunction with the values of the determining variables for every individual—STATA will predict, for each of the 98,732 persons in the sample, the probabilities of belonging to the three different occupational classes by computing the $\hat{p}_{i1}$ from Equation 3.5a and the $\hat{p}_{i2}$ and $\hat{p}_{i3}$ from Equation 3.5b. The mean values of these individual probabilities, computed over all persons and then over the three groups, white, black, and Indian, are denoted, respectively, $\bar{p}_j$, $\bar{p}_{Wj}$, $\bar{p}_{Bj}$, and $\bar{p}_{Ij}$ ($j = 1, 2, 3$). These mean probabilities shown in Table 3.4 (Probabilities as mean of predicted individual probabilities) indicate, for example, that the predicted mean probabilities of all male employees and of white, black, and Indian male employees of being in PMT jobs were, respectively, 41.5%, 41.8%, 29.3%, and 35.1%. It should be emphasized that the predicted mean probabilities

TABLE 3.4

Predicted Probabilities of Whites, Blacks, and Indians Being in
Different Occupational Classes*

	Predicted Probability of Being:		
	UNS	SKL	PMT
Probabilities as mean of predicted individual probabilities			
All persons	17.7	40.8	41.5
White	17.5	40.7	41.8
Black	20.5	50.2	29.3
Indian	25.1	39.8	35.1
Predicted Probabilities at mean values of determining variables			
All persons	14.3	41.8	43.9
White	14.2	41.7	44.1
Black	19.6	52.0	28.4
Indian	22.0	43.3	34.7

*Calculated from the multinomial logit estimates of Table 3.3.
UNS = Unskilled/Semiskilled manual
SKL = Skilled Manual/nonmanual
PMT = Professional/Managerial/Technical

of all persons, and of persons from each ethnic group, of belonging to
the three occupational classes are exactly equal to the corresponding
sample proportions in each class.[50] This is a property of the multinomial logit model, the means of the predicted individual probabilities
of the outcomes are always equal to the sample proportions for the
outcomes.

An alternative way of predicting probabilities is to compute the
mean of the $\widehat{Z}_{ij}$ over all persons and over the white, black, and Indian
groups. If these are denoted, respectively, $\overline{Z}_j$, $\overline{Z}_{Wj}$, $\overline{Z}_{Bj}$ and $\overline{Z}_{Ij}$ then
Equation 3.12a can be used to calculate the probability of the first out-
come for all persons and for each of the groups ($\hat{p}_1$, $\hat{p}_{W1}$, $\hat{p}_{B1}$, and
$\hat{p}_{I1}$), and Equation 3.12b can be used to calculate the probabilities
of the second and third outcomes for all persons and for each of the
groups ($\hat{p}_j$, $\hat{p}_{Wj}$, $\hat{p}_{Bj}$, and $\hat{p}_{Ij}$, ($j = 2, 3$)).[51] These probabilities, shown
in Table 3.4 (Predicted probabilities at mean values of determining
variables), show that at the mean values of the determining variables the predicted probabilities of all, white, black, and Indian male

employees of being in PMT jobs were, respectively, 43.9%, 44.1%, 28.4%, and 34.7%. It is not surprising that these are different from the probabilities reported in the upper part of the table. For example, although both the $\bar{p}_{Wj}$ and the $\hat{p}_{Wj}$ ($j = 1, 2, 3$) purport to measure the overall probability of white persons being in the different occupational classes, they are computed very differently. The $\bar{p}_{Wj}$ are computed as the mean (over the 96,297 white persons) of the estimated individual probabilities, $\hat{p}_{ij}$, where the $\hat{p}_{ij}$ are computed from the values of the determining variables for the individuals. On the other hand, the $\hat{p}_{Wj}$—which bypass the individual probabilities—are computed, instead, from the mean value $\bar{Z}_{Wj}$ of the individual Z_{ij} of the 96,297 white persons in the sample.[52] In general, therefore, for any group g: $\bar{p}_{g1} \neq \hat{p}_{g1}$; $\bar{p}_{g2} \neq \hat{p}_{g2}$ and $\bar{p}_{g3} \neq \hat{p}_{g3}$.

Uncovering Marginal Effects Through Simulations

In the earlier discussion it was emphasized that the direction of change in $\Pr(Y_i = j)$, the probability of observing outcome j for a small change in X_{ir}, could not be inferred from the sign of β_{jr} since $\partial \text{Prob}(Y_i = j)/\partial X_{ir}$ need not have the same sign as β_{jr}. In the context of multinomial logit models, it was only the direction of change in the risk-ratios that could be predicted from the sign of the coefficients. As a consequence, the preceding discussion of the estimation results was cast in terms of such ratios.

Nonetheless, one may often be interested in the underlying probabilities, $\Pr(Y_i = j)$, rather than in the risk-ratios relative to some base outcome, $\Pr(Y_i = j)/\Pr(Y_i = 1)$. In the context of the present application, one might be interested in particular in how the probability of being in a specific occupational class differed between the ethnic groups. Since the coefficient estimates do not directly offer answers to such questions, the alternative is to view such results through the window of simulations. The results of the model can be made transparent by calculating the probabilities of the different outcomes in a variety of hypothetical situations.[53] More specifically, the effects of ethnicity can be analyzed by comparing the probabilities that result when the ethnic dummy variables take different values, *the values of the other variables remaining unchanged between the comparisons.*

Suppose that all the 98,732 persons in the sample were white, so that $IND_i = BLK_i = 0$ for all $i = 1, \ldots, N$. Let p_{ij}^{W} denote the probability of person i being in occupational class j ($j = 1, 2, 3$) in this hypothetical situation. Now suppose that all the 98,732 persons in the sample were black, so that $BLK_i = 1$ for all $i = 1, \ldots, N$ and let p_{ij}^B denote the probability of person i being in occupational class j ($j = 1, 2, 3$) under these hypothetical circumstances.[54] Note that because of the specification of Equation 3.14, the probabilities, p_{ij}^W and p_{ij}^B are the result of all the individuals in the sample having their characteristics evaluated at, respectively, white (α_{jr}) and black ($\alpha_{jr} + \beta_{jr}$) coefficients. They are referred to hereafter as "ethnic" probabilities because they are computed from samples which differ *only* in respect of the ethnicity of the persons comprising the samples. The coefficient estimates, $\hat{\alpha}_{jr}$ and $\hat{\alpha}_{jr} + \hat{\beta}_{jr}$, when applied to Equation 3.5a and 3.5b yield estimates, $\hat{p}_{ij}^W$ and $\hat{p}_{ij}^B$, of the ethnic probabilities.

If $\bar{p}_j^W = \sum_{i=1}^{N} \hat{p}_{ij}^W/N$ and $\bar{p}_j^B = \sum_{i=1}^{N} \hat{p}_{ij}^B/N$ are the respective means of the individual probability estimates (where the latter was computed under the two hypothetical sets of circumstances) then differences between the mean ethnic probabilities $\bar{p}_j^W$ and $\bar{p}_j^B$ are entirely the result of different sets of coefficients (white and black) being applied to a given set of characteristics (that of the N persons in the sample). These differences may, therefore, be attributed entirely to the unequal treatment of persons who, except in their ethnicity, are identical in every respect. More succinctly, they may be attributed to the fact that blacks face "ethnic disadvantage." A similar exercise can be performed for the hypothetical case when everyone in the sample is Indian.

The estimates of these ethnic probabilities are shown in the upper panel (Probabilities as mean of predicted individual probabilities) of Table 3.5 for the three groups, white, black, and Indian. This shows that if everyone in the sample had their characteristics evaluated using the black coefficients $\alpha_{jr} + \beta_{jr}$ instead of having them evaluated at the coefficients relevant to their ethnic group,[55] then 40% *of the total sample* would be in PMT jobs, 39% would be in SKL jobs, and 21% would be in UNS jobs.

An alternative to calculating the ethnic probabilities in this manner would be to compare the probabilities that result when the ethnic variables take their different values and the values of the other variables are held constant in each case to the mean of their sample values. Denote by $\overline{Z}_j^W$ and $\overline{Z}_j^B$ the *mean values* (computed over the 98,732 persons in the sample) of $\widehat{Z}_{ij} = \sum_{r=1}^{R} \hat{\beta}_{jr} X_{ir}$ when everyone in the sample was (treated as) white and everyone in the sample was

TABLE 3.5

Predicted "Ethnic" Probabilities of Whites, Blacks, and Indians
Being in Different Occupational Classes*

	Predicted Probability of Being in Occupations:		
	UNS	SKL	PMT
Probabilities as mean of predicted individual probabilities			
White	17.5	40.7	41.8
Black	20.7	39.5	39.8
Indian	19.3	47.0	33.7
Sample Proportions:			
White	17.5	40.7	41.8
Black	20.5	50.2	29.3
Indian	25.1	39.8	35.1
Predicted Probabilities at mean values of determining variables			
White	14.2	41.6	44.2
Black	19.3	39.7	41.0
Indian	16.6	50.8	32.6

*Calculated from the multinomial logit estimates of Table 3.3.
UNS = Unskilled/Semiskilled manual
SKL = Skilled Manual/nonmanual
PMT = Professional/Managerial/Technical

(treated as) black.[56] Then the white and black probabilities of being in the different occupational classes (respectively denoted, $\hat{p}_j^W$ and $\hat{p}_j^B$, $j = 1, 2, 3$) are calculated using $\overline{Z}_j^W$ and $\overline{Z}_j^B$ in Equation 3.12a for the first outcome and in Equation 3.12b for the second and third outcomes. The values of these estimated probabilities are shown for whites, blacks, and Indians in Table 3.5. (Predicted probabilities at mean values of determining variables) When the probabilities are computed from the average characteristics of the sample, a larger proportion of whites and blacks (and a smaller proportion of Indians) are predicted to be in the PMT class ($\hat{p}_3^W = 44.2\%$, $\hat{p}_3^B = 41.0\%$, $\hat{p}_3^I = 32.6\%$) than were predicted when the probabilities were computed as the mean of the individual probabilities ($\bar{p}_3^W = 41.8\%$, $\bar{p}_3^B = 39.8\%$, $\bar{p}_3^I = 33.7\%$). This is a consequence of Indians having a higher than average endowment of higher educational qualifications (see Table 3.1) which is a major determinant of success in obtaining

PMT jobs. As a consequence, when everyone is reduced to the sample mean, Indians "lose out" and other groups "gain."

However, as Table 3.5 (Sample Proportions) shows, only 29% of black male employees were *actually* in PMT jobs. The sample proportions of blacks and whites in occupation class j, denoted, respectively, s_j^B and s_j^W, will, in general, be different from the ethnic probabilities, $\bar{p}_j^B$ and $\bar{p}_j^W$ (and from $\hat{p}_j^B$ and $\hat{p}_j^W$). This reflects the fact that blacks and whites differ not just in terms of how they are treated but also in terms of their characteristics. The fact that the sample proportion of blacks in PMT jobs ($s_3^B = 29.3\%$) is less than their predicted ethnic probability ($\bar{p}_3^B = 39.8\%$ or $\hat{p}_3^B = 41.0\%$) of being in such jobs is due to the fact that, *relative to the sample in its entirety*, the characteristics of blacks are less suited to PMT jobs. In other words, blacks, relative to whites, face, with respect to PMT jobs, *both* ethnic and characteristics disadvantage. The total of these separate disadvantages is referred to as the *overall disadvantage*.

Measuring Occupational Disadvantage

A measure of the *ethnic disadvantage* experienced, on average, by blacks vis-à-vis whites, with respect to occupational class j is λ_j^B, where

$$\lambda_j^B = \bar{p}_j^B / \bar{p}_j^W. \tag{3.15}$$

If the two probabilities are equal, then $\lambda_j^B = 1$ and there is no ethnic disadvantage. However, if $\bar{p}_j^B < \bar{p}_j^W$, then $\lambda_j^B < 1$ and there is black ethnic disadvantage[57] for occupational class j, with the size of this disadvantage being greater the further λ_j^B is from 1.

A measure of the overall disadvantage experienced, on average, by blacks vis-à-vis whites, with respect to occupational class j is μ_j^B, where

$$\mu_j^B = s_j^B / s_j^W. \tag{3.16}$$

If the two sample proportions are equal, then $\mu_j^B = 1$ and there is no overall disadvantage. However, if $s_j^B < s_j^W$, then $\mu_j^B < 1$ and there is black overall disadvantage[58] for occupational class j, with the size of this disadvantage being greater the further μ_j^B is from 1.

A measure of the characteristics disadvantage experienced, on average, by blacks vis-à-vis whites with respect to occupational class j is

δ_j^B, where δ_j^B is the ratio of the overall disadvantage to the ethnic disadvantage,

$$\delta_j^B = \mu_j^B / \lambda_j^B.$$

If $\delta_j^B = 1$, then blacks do not face a characteristics disadvantage since $\mu_j^B = \lambda_j^B$. In this case the ratio of the black and white sample proportions is equal to the corresponding ratio of the ethnic probabilities. If the latter ratio is less than 1, then it is entirely due to the identical characteristics of blacks and whites being evaluated more favorably for the latter than for the former. However, if $\delta_j^B < 1$, then there is a characteristic disadvantage, and the size of this disadvantage is greater the further δ_j^B is from 1. Blacks are penalized for their characteristics since then $\mu_j^B < \lambda_j^B$. They lose out by having inferior characteristics, in addition to (possibly) having these characteristics evaluated less favorably than they would have been had they been the characteristics of whites. Lastly, if $\delta_j^B > 1$, then blacks enjoy an characteristics advantage since then $\mu_j^B > \lambda_j^B$. Blacks draw the sting from possible unfair treatment by acquiring superior characteristics. Indeed, if the value of δ_j^B was sufficiently large then it would be possible for $\mu_j > 1$, even though $\lambda_j < 1$. In other words, the superior characteristics of blacks would more than neutralize any unfair treatment that they might experience. Table 3.6 shows estimates of the three disadvantages for each of the three ethnic groups.

The figures under the column "Ethnic Disadvantage" in Table 3.6 indicate that if Indians and whites had been assigned a common set of characteristics (namely, the characteristics of the sample as a whole) then the probability of Indians being in the PMT class would have been 81% of the corresponding white probability. A similar exercise for blacks would have yielded a value of 95%. However, the fact that Indians and blacks, as distinct groups, had characteristics that were different from those of the sample, considered in its entirety, meant that the sample proportions of Indians and blacks in the PMT class were, respectively, 84% and 70% of the corresponding white proportion. This rise for Indians from the ethnic 81% to the sample 84%, and the corresponding fall for blacks from the ethnic 95% to the sample 70%, could be attributed, respectively, to "superior" Indian characteristics and "inferior" black characteristics. Table 3.6 shows that, with regard to PMT employment, Indians had a characteristics advantage of 4%, but blacks had a characteristics disadvantage of 15%, over

TABLE 3.6

Estimates of Ethnic, Characteristics, and Overall Disadvantage*
Faced by Blacks and Indians Relative to Whites

	Ethnic Disadvantage (%)	Characteristics Disadvantage (%)	Overall Disadvantage (%)
Disadvantage calculated from probabilities as mean of predicted individual probabilities			
Black/White			
$j = 1$ (SKL)	0.97	1.27	1.23
$j = 2$ (PMT)	0.95	0.74	0.70
Indian/White			
$j = 1$ (SKL)	1.15	0.85	0.98
$j = 2$ (PMT)	0.81	1.04	0.84
Disadvantage calculated from predicted probabilities at mean values of determining variables			
Black/White			
$j = 1$ (SKL)	0.95	1.26	1.23
$j = 2$ (PMT)	0.93	0.75	0.70
Indian/White			
$j = 1$ (SKL)	1.22	0.80	0.98
$j = 2$ (PMT)	0.73	1.15	0.84

*Advantage if value > 1
Calculated from the figures of Tables 3.5
UNS = Unskilled/Semiskilled manual
SKL = Skilled Manual/nonmanual
PMT = Professional/Managerial/Technical

their white counterparts. With respect to the SKL class, Indians had a characteristics disadvantage, relative to whites, of 15% while blacks had a characteristics advantage of 27% over their white counterparts.

One could also have defined ethnic disadvantage as $\hat{p}_j^B / \hat{p}_j^W$ —that is, as the ratio of the probabilities (shown in the lower panel of Table 3.5) obtained by setting the values of all except the ethnic variables to their sample means. The characteristics disadvantage is now recomputed as the ratio of the overall disadvantage[59] and the *new* estimate of ethnic disadvantage. The new values of ethnic and characteristics disadvantage (as well as the unchanged value of overall disadvantage) are shown in Table 3.6.

**Conditional Logit and
the Independence of Irrelevant Alternatives**

In a conditional logit model the outcome probabilities $\Pr(Y_i = j)$ depend only upon the choice *attributes* and not upon the *characteristics* of the individuals making the choices. In other words, the conditional logit model is defined by Equation 3.3 but with the caveat that now, with the $\beta_{jr} = 0$,

$$Z_{ij} = \sum_{s=1}^{S} \gamma_{is} W_{js}. \tag{3.17}$$

Apart from this, the model is essentially the same as the multinomial logit model. Usually the reason for omitting the characteristics of the individuals is that data on individuals are not available. Consequently, the conditional logit model can be usefully thought of as a collective of persons—hereafter, referred to as the "population"—making choices between alternatives with different attributes—for example, commuters choosing between modes of travel or consumers choosing between supermarkets. With this mind, the subscript i relating to individuals is dropped in the exposition that follows. The dependent variable $Y = j$ relates to the choice made by the population and the coefficient γ_s refers to the weight attached by the population to the sth attribute. As in the multinomial model, the risk-ratios of any two alternatives j and k in the conditional logit model are independent of the other available alternatives,

$$\log\left(\frac{\Pr(Y = j)}{\Pr(Y = k)}\right) = Z_j - Z_k = \sum_{s=1}^{S}(W_{js} - W_{ks})\gamma_s. \tag{3.18}$$

In Equation 3.18 the log risk-ratio depends solely on the (differences between) attributes associated with j and k and is independent of the attributes associated with any other alternative.[60] This property, known as the *Independence of Irrelevant Alternatives (IIA)*, is both the principal strength and the principal weakness of such models.[61] It is a strength because it allows the introduction of new alternatives without the need to re-estimate the model. For example, suppose a $(M+1)$th alternative is introduced with attributes $W_{M+1,s}$ $s = 1, \ldots, S$. The estimated Z-value associated with this new alternative is: $\widehat{Z}_{M+1} = \sum_{s=1}^{S} \hat{\gamma}_s W_{M+1,s}$, where the coefficient estimates $\hat{\gamma}_s$ *have already been obtained*. All that the introduction of a new alternative requires is that a new term $\widehat{Z}_{M+1}$ to be added to the numerator in Equation 3.3 and for the probabilities to be recomputed. The reason that the existing coefficient estimates will also serve when the set of alternatives is expanded is that the addition of an alternative cannot change the relative risk with which existing alternatives are chosen.

The percentage change in the probability of choosing a particular alternative, given the introduction of a new alternative is easily computed as

$$\frac{\Pr(Y^{M+1} = j) - \Pr(Y^M = j)}{\Pr(Y^M = j)} = \left(\frac{\exp(Z_j)}{\sum_{j=1}^{M+1} \exp(Z_j)} \frac{\sum_{j=1}^{M} \exp(Z_j)}{\exp(Z_j)} \right) - 1$$

$$= \frac{-\exp(Z_{M+1})}{\sum_{j=1}^{M+1} \exp(Z_j)}$$

$$= -\Pr(Y^{M+1} = M+1),$$

where $\Pr(Y^{M+1} = j)$ and $\Pr(Y^M = j)$ refer to the probability of choosing alternative j when there are, respectively, $M+1$ and M alternatives available. As a consequence of introducing an additional alternative the probabilities of choosing the existing alternatives will all fall by the *same* percentage, which is equal to the probability of choosing the new alternative.

However, this property of a uniform percentage drop in all existing probabilities in the face of a new alternative is also a weakness because it implies that the cross-elasticity of demand for an existing alternative with respect to the new alternative is uniform across the alternatives.[62] For this to be valid, the alternatives must be viewed as completely distinct and independent. The fact that they are so viewed, which gives rise to IIA, stems from the assumption that the errors are independently and identically distributed.

How debilitating this limitation of the model is depends on the nature of the problem being analyzed. In a model of occupational choice, for example, where the alternatives might be thought to be well-defined and relatively immutable, the issue of introducing new alternatives might not be a serious one.[63] On the other hand, with commuters' choice of travel mode the introduction of new modes of transport—new train lines, bus lanes, congestion charges on cars, etc.—must be considered a serious possibility.

A classic example of problems caused by the IIA assumption is the "red bus-blue bus" problem. Suppose that commuters have three choices for travel to work, car, train, or bus. Suppose, parenthetically, that all the buses are painted red and that the logit predictions of $\Pr(Y = j)$ are

- Car: 55.4
- Bus (Red): 23.0
- Train: 21.6

The risk-ratio between car and bus travel is $55.4/23.0 = 2.4$. Now the bus company decides to paint half its fleet blue. It is reasonable to expect that this purely cosmetic change would leave commuters' choices unaffected and that the same proportions, as shown above, would continue to travel by car, bus and train. However, since the model cannot distinguish between a spurious and a genuine new alternative, with a "fourth" alternative the predicted logit probabilities are[64]

- Car: 45.1
- Bus (Red): 18.8
- Bus (Blue): 18.4
- Train: 17.7

Note that the probability of each of the existing alternatives, car, (red) bus, train has fallen by 18.4%, the probability of travel by blue bus. The risk-ratios between car and red bus (and between train and red bus) travel remain unchanged but, in order to accommodate this, some commuters who earlier travelled by car, train, or red bus *have to change to travel by blue bus.* As a consequence, by simply painting half its fleet blue, the bus company could increase the take-up of bus travel from 23% of all commuters to 37.2%!

A more substantive example might be found in politics. Suppose a political party X is in competition with two other parties, Y and Z. Using a multinomial logit model one can estimate the likelihood of voters voting for Z as a ratio of the likelihood of voting for party X. Suppose now party X splits into two parties, $X1$ and $X2$, which hold similar (say, left-wing) views and which are generally regarded by voters as being not substantially different from each other. However, since the model cannot distinguish between a spurious and a genuine new alternative, with a "fourth" party the predicted logit probabilities of voting for a left-wing party would be predicted to rise substantially. This then is the great weakness of conditional logit: under IIA, the model offers no protection against spurious share inflation through the introduction of similar, or identical, alternatives.

Parenthetically one should note a further limitation that arises when one considers compound choices—for example, the simultaneous choice of travel time and travel mode. Then, as Domencich and McFadden (1996) show, the joint probability of choosing time t and mode j, $\Pr(t \cap j)$, can be written as the product of: (the probability of choosing mode j, from the available choice of modes, at time t) and (the probability of choosing time t from the available choice of times). In other words, an implication of IIA is that the utility function embodying a compound choice must be additively separable from the individual choices, $U(t, j) = \phi(j) + \psi(t)$.

Alternatives to the Logit Model

Before considering alternatives to the logit model it is important to examine whether the assumption of IIA is or is not valid. Suppose one believes that a subset of the choice set is irrelevant. Eliminating it from the choice set should not significantly alter the coefficient estimates. This thinking lies behind the Hausman and McFadden (1984) test. If $\hat{\gamma}_1$ is the vector of coefficient estimates based on the restricted set of choices, $\hat{\gamma}_0$ is the vector of coefficient estimates based on the full set of choices, and $\widehat{V}_1 - \widehat{V}_0$ are the respective estimates of their covariance matrices, then the statistic:

$$(\hat{\gamma}_1 - \hat{\gamma}_0)'(\widehat{V}_1 - \widehat{V}_0)(\hat{\gamma}_1 - \hat{\gamma}_0)$$

is distributed as $\chi^2(S)$ under the null hypothesis that the restrictions of IIA are validly imposed.

If this null hypothesis is not accepted, then one possibility is to specify the random utility function of Equation 3.1 as a *multivariate probit* model,

$$U_j = \sum_{s=1}^{S} \gamma_s W_{js} + \varepsilon_j \quad \varepsilon_j \sim N(0, \Sigma),$$

where with a nonscalar covariance matrix Σ the errors need no longer be independent. The problem with this model is the practical one of computing the multinormal integration and estimation of the unrestricted covariance matrix (Greene, 2000, p. 865).

Another way of relaxing the IIA restriction is to group the alternatives in such a way that the variances differ across the groups, but are the same within each group. Then the IIA assumption is relaxed between groups of alternatives but is maintained within each group. This is the method of *nested logit*. This topic is not pursued here but details and an application can be found in Greene (1995).

4. PROGRAM LISTINGS

Introduction

This chapter contains the computer listings of the *STATA* programs that generated the results discussed in the earlier chapters. The point was made in the introductory chapter—and it bears repeating—that there are at least four well-known, highly regarded pieces of software which, among other things, handle problems of the kind discussed in this monograph: *SAS; SPSS v10.0* (A good introduction to the use of SAS and SPSS procedures in the analysis of events with ordered and unordered outcomes is provided by the Stat/Math Center at the Indiana University[65]); *LIMDEP* (see Greene, 1995); and *STATA* (see STATA, 1999). It just so happens that I am more familiar with STATA than with the others. For this reason, and this reason alone, the programs of this chapter are written in STATA code.

The next section contains the program listings of the ordered logit and probit models of Chapter 2, and section 3 contains the program listings of the multinomial logit model of Chapter 3. The programs are liberally interspersed with comments. These comments try to fulfill three aims:

- explain what a particular piece of program code is going to do (or has done);
- relate the piece of code to a specific table, referenced in the earlier chapters, so that the reader can see how the results were generated
- relate the piece of code to specific equations, referenced in the earlier chapters, so that the reader can see the methodology employed to generate a particular result

Ordered Probit and Logit Programs

version 6.0 /* Using STATA version 6.0 */;
   use C:\SAGE\NI.dta /* Reading data that are in STATA
      format: 58 vars, 13,164 observations */;
   #delimit; /* Commands will be terminated by; */;
   /* TITLE: ORDERED LOGIT AND PROBIT USING
      DEPRIVATION EXAMPLE */;
   gen pnum = _n; /* Each person is given a number */;
   /* y is dependent variable for ordered logit
   $y = 1$ is not deprived; $y = 2$ is mildly deprived; $y = 3$ is
      severely deprived */;
   tabulate y rc, col; tabulate y sex, col;
   /* Oprobit equation is being estimated on entire subsample:
      Table 2.2 */;
   oprobit y sex ct age age2 ret inac ue highed mided hnum
      snpar ard dwn crk ant col arm ban dry frm, table;
   predict p1 p2 p3; /* Predicted probabilities are stored
      in p1, p2, p3 */;
   sort pnum; /* Observations are sorted by person number in
      ascending order */;
   list pnum p1 p2 p3 in 1/25, noobs; /* Predicted probabilities
      of first 25 persons are listed with person number: Table 2.4 */;
   summarize p1 p2 p3, detail; /* Predicted probabilities are
      summarized with detail: Table 2.6 */;
   predict z, xb; /* The value of Z for each person is computed */;
   summarize z, mean; gen zm = r(mean); /* The mean of Z is
      computed and stored in zm */;
   /* b[_cut1] and b[_cut2] store the estimated values of the cutoff
      points, $\hat{\delta}_1$ and $\hat{\delta}_2$ */;
   gen p1 = normprob(_b[_cut1] − z); gen p2 = normprob (_b[_cut2]
      − z) − normprob(_b[_cut1] − z); gen p3 = 1 − normprob
      (_b[_cut2] − z);
   /* For each person, p1, p2 and p3 are predicted the probabilities,
      calculated using Equations 2.5a to 2.5c.
   They are the same as the predicted p1–p3 calculated earlier,
      above */;
   gen q1 = normprob(_b[_cut1] − zm); gen q2 = normprob(_b[_cut2]

$- zm) - normprob(\_b[\_cut1] - zm)$; gen $q3 = 1 -$ normprob
$(\_b[\_cut2] - zm)$;
/* q1, q2 and q3 are the predicted probabilities, calculated
setting values to sample means: Equations 2.18a to 2.18c */;
summarize p1 p2 p3; summarize q1 q2 q3; /* Comparing
probabilities: Table 2.8 */;
drop p1 p2 p3 q1 q2 q3; /* Releasing variable names for
subsequent use */;
/* Calculating Marginal Effects: Continuous Variables (Age) */;
gen a1 = normd(_b[_cut1] - z)*_b[age];
gen a2 = (normd(_b[_cut2] - z) - normd(_b[_cut1] - z))*_b[age];
gen a3 = -1*normd(_b[_cut2] - z)*_b[age];
/* a1-a3 are marginal effects for AGE_i calculated for each
individual: Equations 2.13a to 2.13c */;
gen am1 = normd(_b[_cut1] - zm)*_b[age];
gen am2 = (normd(_b[_cut2] - zm) - normd(_b[_cut1] - zm))
*_b[age];
gen am3 = -1*normd(_b[_cut2] - zm)*_b[age];
/* am1-am3 are marginal effects for AGE setting variable values to
sample mean */;
gen b1 = normd(_b[_cut1] - z)*_b[age2];
gen b2 = (normd(_b[_cut2] - z) - normd(_b[_cut1] - z))*_b[age2];
gen b3 = -1*normd(_b[_cut2] - z)*_b[age2];
/* a1-a3 are marginal effects for AGE_i^2 calculated for each
individual: Equations 2.13a to 2.13c */;
gen bm1 = normd(_b[_cut1] - zm)*_b[age2];
gen bm2 = (normd(_b[_cut2] - zm) - normd(_b[_cut1] - zm))
*_b[age2];
gen bm3 = -1*normd(_b[_cut2] - zm)*_b[age2];
/* am1-am3 are marginal effects for AGE^2 setting variable
values to sample mean */;
gen c1 = a1 + b1; gen c2 = a2 + b2; gen
c3 = a3 + b3; gen cm1 = am1 + bm1; gen cm2 = am2 + bm2;
gen cm3 = am3 + bm3;
/* Add effects of AGE and AGE^2 */;
summarize c1 c2 c3; summarize cm1 cm2 cm3; /* Table 2.10 */;
drop z zm c1 a1 b1 c2 a2 b2 c3 a3 b3 cm1 am1 bm1 cm2 am2
bm2 cm3 am3 bm3;

/* Calculating Marginal Effects: Dummy Variables (Religion) */;
gen cto = ct; /* Saving original values */;
replace ct = 1; /* Everyone is Catholic */;
predict p1 p2 p3; /* Predicted probabilities for each person
    when $CT_i = 1$ */;
predict z, xb; summarize z; gen zm = r(mean); /* The mean of
    Z is computed and stored in zm */;
gen q1 = normprob($_b[_cut1]$ − zm); gen q2 = normprob
    ($_b[_cut2]$ − zm) − normprob($_b[_cut1]$ − zm); gen
    q3 = 1 − normprob($_b[_cut2]$ − zm);
/* q1, q2 and q3 are the predicted probabilities, calculated
    setting nonreligion values to sample means */;
summarize p1 p2 p3; summarize q1 q2 q3; /* Table 2.12 */;
drop p1 p2 p3 q1 q2 q3 z zm;
replace ct = 0; /* Everyone is Protestant */;
predict p1 p2 p3; /* Predicted probabilities for each person
    when $CT_i = 0$ */;
predict z, xb; summarize z; gen zm = r(mean); /* The mean
    of Z is computed and stored in zm */;
gen q1 = normprob($_b[_cut1]$ − zm); gen q2 = normprob
    ($_b[_cut2]$ − zm) − normprob($_b[_cut1]$ − zm);
    gen q3 = 1 − normprob($_b[_cut2]$ − zm);
/* q1, q2 and q3 are the predicted probabilities, calculated
    setting non-religion values to sample means */;
summarize p1 p2 p3; summarize q1 q2 q3; /* Table 2.12 */;
drop p1 p2 p3 q1 q2 q3 z zm;
replace ct = cto; /* Restoring original values */;
/* Ologit equation is being estimated on entire subsample:
    Table 2.1 */;
ologit y sex ct age age2 ret inac ue highed mided hnum snpar
    ard dwn crk ant col arm ban dry frm, table;
predict p1 p2 p3; /* Predicted probabilities are stored
    in p1, p2, p3 */;
sort pnum; /* Observations are sorted by person number in
ascending order */;
list pnum p1 p2 p3 in 1/25, noobs; /* Predicted probabilities
    of first 25 persons are listed, with person number: Table 2.4 */;
summarize p1 p2 p3, detail; /* Predicted probabilities are

summarized with detail: Table 2.6 */;
predict z, xb; /* The value of Z for each person is computed */;
summarize z, mean; gen zm = r(mean); /* The mean of
 Z is computed and stored in zm */;
/* b[_cut1] and b[_cut2] store the estimated values of the
 cutoff points, $\hat{\delta}_1$ and $\hat{\delta}_2$ */;
gen p1 = 1/(1 + exp(z − _b[_cut1])); gen p2 = 1/(1 + exp(z −
 _b[_cut2])) − 1/(1+exp(z − _b[_cut1])); gen p3 = 1−
 1/(1 + exp(z − _b[_cut2]));
/* For each person, p1, p2 and p3 are predicted the probabilities,
 calculated using equations 2.5a to 2.5c.
They are the same as the predicted p1–p3 calculated earlier,
 above */;   gen q1 = 1/(1 + exp(zm − _b[_cut1])); gen
 q2 = 1/(1 + exp(zm − _b[_cut2])) − 1/(1 + exp(zm − _b
 [_cut1]));
 gen q3 = 1 − 1/(1 + exp(zm − _b[_cut2]));
/* q1, q2 and q3 are the predicted probabilities, calculated
 setting values to sample means: Equations 2.18a to 2.18c */;.
summarize p1 p2 p3; summarize q1 q2 q3; /* Comparing
 probabilities: Table 2.7 */;
drop p1 p2 p3 q1 q2 q3; /* Releasing variable names for
 subsequent use */;
/* Calculating Marginal Effects: Continuous Variables
 (Age) */;
gen lden1 = 1/(1 + exp(z − _b[_cut1]));
gen lden2 = 1/(1 + exp(z − _b[_cut2]));
gen lden3 = 1/(1 + exp(z − _b[_cut2]));
/* Logit probabilities are calculated: Equations 2.9a to 2.9c */;
gen a1 = lden1*(1 − lden1)*_b[age];
gen b1 = lden1*(1 − lden1)*_b[age2];
gen c1 = a1+b1;
gen a2 = (lden2*(1 − lden2) − lden1*
 (1 − lden1))*_b[age];
gen b2 = (lden2*(1 − lden2) − lden1*
 (1 − lden1))*_b[age2];
gen c2 = a2+b2;
gen a3 = −lden3*(1 − lden3)*_b[age];
gen b3 = −lden3*(1 − lden3)*_b[age2];

```
gen c3 = a3 + b3;
/* a1–a3 are marginal effects for AGE_i calculated for each
   individual: Equations 2.12a to 2.12c */;
/* b1–b3 are marginal effects for AGE_i^2 calculated for each
   individual: Equations 2.12a to 2.12c */;
/* c1–c3 is the sum of effects */;
replace lden1 = 1/(1 + exp(zm − _b[_cut1]));
replace lden2 = 1/(1 + exp(zm − _b[_cut2]));
replace lden3 = 1/(1 + exp(zm − _b[_cut2]));
/* Logit probabilities are calculated at mean */;
gen am1 = lden1*(1 − lden1)*_b[age];
gen bm1 = lden1*(1 − lden1)*_b[age2];
gen cm1 = am1 + bm1;
gen am2 = (lden2*(1 − lden2) − lden1* (1 − lden1))*_b[age];
gen bm2 = (lden2*(1 − lden2) − lden1*(1 − lden1))*_b[age2];
gen cm2 = am2 + bm2;
gen am3 = −lden3*(1 − lden3)*_b[age];
gen bm3 = −lden3*(1 − lden3)*_b[age2];
gen cm3 = am3 + bm3;
/* a1–a3 are marginal effects for AGE_i calculated
   at sample means */;
/* b1–b3 are marginal effects for AGE_i^2 calculated
   at sample means */;
/*c1–c3 is the sum of effects */;
summarize c1 c2 c3; summarize cm1 cm2 cm3; /* Table 2.9 */;
drop z zm c1 a1 b1 c2 a2 b2 c3 a3 b3 cm1 am1 bm1 cm2 am2
   bm2 cm3 am3 bm3;
/* Calculating Marginal Effects: Dummy Variables (Religion) */;
replace ct = 1; /* Everyone is Catholic */;
predict p1 p2 p3; /* Predicted probabilities for each person
   when CT_i=1 */;
predict z, xb; summarize z; gen zm = r(mean); /* The mean
   of Z is computed and stored in zm */;
gen q1 = 1/(1 + exp(zm − _b[_cut1])); gen q2 = 1/(1 + exp
   (zm − _b[_cut2])) − 1/(1 + exp(zm − _b[_cut1])); gen q3
   = 1 − 1/(1+exp(zm − _b[_cut2]));
/* q1, q2 and q3 are the predicted probabilities, calculated
   setting nonreligion values to sample means */;
```

```
summarize p1 p2 p3; summarize q1 q2 q3; /* Table 2.11 */;
drop p1 p2 p3 q1 q2 q3 z zm;
replace ct = 0; /* Everyone is Protestant */;
predict p1 p2 p3; /* Predicted probabilities for each person
   when CT_i=0 */;
predict z, xb; summarize z; gen zm = r(mean); /* The mean
   of Z is computed and stored in zm */;
gen q1 = 1/(1 + exp(zm - _b[_cut1])); gen q2 = 1/(1 + exp
   (zm - _b[_cut2])) - 1/(1 + exp(zm - _b[_cut1]));
   gen q3 = 1 - 1/(1+exp(zm - _b[_cut2]));
/* q1, q2 and q3 are the predicted probabilities, calculated
   setting nonreligion values to sample means */;
summarize p1 p2 p3; summarize q1 q2 q3; /* Table 2.11 */;
drop p1 p2 p3 q1 q2 q3 z zm;
replace ct = cto; /* Restoring original values */;
/* Ologit equation is being estimated on Protestant subsample */;
ologit y sex age age2 ret inac ue highed mided hnum snpar ard
   dwn crk ant col arm ban dry frm if ct == 0, table;
lrtest,saving(0); /* Likelihood value saved for LR test */;
ologit y sex age age2 ret inac ue highed mided hnum snpar ard
   crk ant col arm ban frm if ct == 0, table;
lrtest; /* LR test on coefficients on dwn & dry jointly zero */;
/* Predictions are being made for Protestants using Protestant
   coefficients and means computed */;
predict ppa1 ppa2 ppa3 if ct == 0;
egen mppa1 = mean(ppa1); egen mppa2 = mean(ppa2); egen
   mppa3 = mean(ppa3);
predict pps1 pps2 pps3 if ct == 0 & snpar == 1;
egen mpps1 = mean(pps1); egen mpps2 = mean(pps2); egen
   mpps3 = mean(pps3);
predict ppr1 ppr2 ppr3 if ct == 0 & ret == 1;
egen mppr1 = mean(ppr1); egen mppr2 = mean(ppr2); egen
   mppr3 = mean(ppr3);
predict ppi1 ppi2 ppi3 if ct == 0 & inac == 1;
egen mppi1 = mean(ppi1); egen mppi2 = mean(ppi2); egen
   mppi3 = mean(ppi3);
predict ppu1 ppu2 ppu3 if ct == 0 & ue == 1;
egen mppu1 = mean(ppu1); egen mppu2 = mean(ppu2); egen
   mppu3 = mean(ppu3);
```

```
predict ppl1 ppl2 ppl3 if ct == 0 & resnum == 1;
egen mppl1 = mean(ppl1); egen mppl2 = mean(ppl2); egen
    mppl3 = mean(ppl3);
/* Ologit equation is being estimated on Catholic subsample */;
ologit y sex age age2 ret inac ue highed mided hnum snpar ard
    dwn crk ant col arm ban dry frm if ct == 1, table;
lrtest, saving(0); /* Likelihood value saved for LR test */;
ologit y sex age age2 ret inac ue highed mided hnum snpar ard
    crk ant col arm ban frm if ct == 1, table;
lrtest; /* LR test on coefficients on dwn & dry jointly zero */;
/* Predictions are being made for Catholics using Catholic
    coefficients and means computed */;
predict cca1 cca2 cca3 if ct == 1;
egen mcca1 = mean(cca1); egen mcca2 = mean(cca2); egen
    mcca3 = mean(cca3);
predict ccs1 ccs2 ccs3 if ct == 1 & snpar == 1;
egen mccs1 = mean(ccs1); egen mccs2 = mean(ccs2); egen
    mccs3 = mean(ccs3);
predict ccr1 ccr2 ccr3 if ct == 1 & ret == 1;
egen mccr1 = mean(ccr1); egen mccr2 = mean(ccr2); egen
    mccr3 = mean(ccr3);
predict cci1 cci2 cci3 if ct == 1 & inac == 1;
egen mcci1 = mean(cci1); egen mcci2 = mean(cci2); egen
    mcci3 = mean(cci3);
predict ccu1 ccu2 ccu3 if ct == 1 & ue == 1;
egen mccu1 = mean(ccu1); egen mccu2 = mean(ccu2); egen
    mccu3 = mean(ccu3);
predict ccl1 ccl2 ccl3 if ct == 1 & resnum == 1;
egen mccl1 = mean(ccl1); egen mccl2 = mean(ccl2); egen
    mccl3 = mean(ccl3);
/* Predictions are being made for Protestants using Catholic
    coefficients and means computed */;
predict pca1 pca2 pca3 if ct == 0;
egen mpca1 = mean(pca1); egen mpca2 = mean(pca2); egen
    mpca3 = mean(pca3);
predict pcs1 pcs2 pcs3 if ct == 0 & snpar == 1;
egen mpcs1 = mean(pcs1); egen mpcs2 = mean(pcs2); egen
    mpcs3 = mean(pcs3);
predict pcr1 pcr2 pcr3 if ct == 0 & ret == 1;
```

```
egen mpcr1 = mean(pcr1); egen mpcr2 = mean(pcr2); egen
    mpcr3 = mean(pcr3);
predict pci1 pci2 pci3 if ct == 0 & inac == 1;
egen mpci1 = mean(pci1); egen mpci2 = mean(pci2); egen
    mpci3 = mean(pci3);
predict pcu1 pcu2 pcu3 if ct == 0 & ue == 1;
egen mpcu1 = mean(pcu1); egen mpcu2 = mean(pcu2); egen
    mpcu3 = mean(pcu3);
predict pcl1 pcl2 pcl3 if ct == 0 & resnum == 1;
egen mpcl1 = mean(pcl1); egen mpcl2 = mean(pcl2); egen
    mpcl3 = mean(pcl3);
/* Now percentage contributions will be calculated */;
gen Aa1 = ((mppa1 – mpca1)/(mppa1 – mcca1))*100;
gen Aa2 = ((mppa2 – mpca2)/(mppa2 – mcca2))*100;
gen Aa3 = ((mppa3 – mpca3)/(mppa3 – mcca3))*100;
gen As1 = ((mpps1 – mpcs1)/(mpps1 – mccs1))*100;
gen As2 = ((mpps2 – mpcs2)/(mpps2 – mccs2))*100;
gen As3 = ((mpps3 – mpcs3)/(mpps3 – mccs3))*100;
gen Ar1 = ((mppr1 – mpcr1)/(mppr1 – mccr1))*100;
gen Ar2 = ((mppr2 – mpcr2)/(mppr2 – mccr2))*100;
gen Ar3 = ((mppr3 – mpcr3)/(mppr3 – mccr3))*100;
gen Ai1 = ((mppi1 – mpci1)/(mppi1 – mcci1))*100;
gen Ai2 = ((mppi2 – mpci2)/(mppi2 – mcci2))*100;
gen Ai3 = ((mppi3 – mpci3)/(mppi3 – mcci3))*100;
gen Au1 = ((mppu1 – mpcu1)/(mppu1 – mccu1))*100;
gen Au2 = ((mppu2 – mpcu2)/(mppu2 – mccu2))*100;
gen Au3 = ((mppu3 – mpcu3)/(mppu3 – mccu3))*100;
gen Al1 = ((mppl1 – mpcl1)/(mppl1 – mccl1))*100;
gen Al2 = ((mppl2 – mpcl2)/(mppl2 – mccl2))*100;
gen Al3 = ((mppl3 – mpcl3)/(mppl3-mccl3))*100;
summarize
ppa1 ppa2 ppa3 pca1 pca2 pca3 cca1 cca2 cca3
pps1 pps2 pps3 pcs1 pcs2 pcs3 ccs1 ccs2 ccs3
ppr1 ppr2 ppr3 pcr1 pcr2 pcr3 ccr1 ccr2 ccr3
ppi1 ppi2 ppi3 pci1 pci2 pci3 cci1 cci2 cci3
ppu1 ppu2 ppu3 pcu1 pcu2 pcu3 ccu1 ccu2 ccu3
ppl1 ppl2 ppl3 pcl1 pcl2 pcl3 ccl1 ccl2 ccl3;
/* Table 2.16 */;
summarize Aa1 Aa2 Aa3 As1 As2 As3 Ar1 Ar2 Ar3
Ai1 Ai2 Ai3 Au1 Au2 Au3 Al1 Al2 Al3; /* Table 2.17 */;
```

/* End of program */;

Multinomial Logit Programs

version 6.0 /* Using STATA version 6.0 */
    use c:\SAGE\GB.dta /* Reading data which are
      in STATA format: 24 vars 98732 observations:
    data are for white, black and Indian full-time male employees,
      25–45 years of age */;
    #delimit;
    /* TITLE: MULTINOMIAL LOGIT USING OCCUPATIONAL
      CLASS EXAMPLE */;   gen pnum = _n; /* Every person is
      assigned a number */;
    /* y is dependent variable for multinomial logit
      $y = 1$ if person is in unskilled/semiskilled manual occupation
      $y = 2$ if person is in skilled manual/nonmanual occupation
      $y = 3$ if person is in professional/managerial/technical
        occupation */;
    /* Tabulating y for all men; white men; black men; Indian
      men */;
    tab y; tab y if blk == 0 & ind == 0; tab y if
      blk == 1; tab y if ind == 1;
    /* Now estimating multinomial logit equation, Table 3.2: see
      Equation 3.14 for equation specification.
    Note: base(1) below sets outcome 1 as base value */;
    mlogit y age age2 blk blkovs ind indovs ovsbn north indnth
      blknth south indsth blksth highed indhe blkhe mided
      indme blkme subbus subsci bush scih, base(1);
    lrtest, saving(0); /* Saving log-likelihood value for LR test */;
    /* Defining zero constraints:
    blk, blkovs, dropped from equation for $y = 2$;
    ind, blknth, scih, bush, subsci dropped from equation for $y = 3$;
    indme, indhe, blkme dropped from both equations */;
    constraint define 1 [2]blkovs = 0; constraint define 2 [3]blk = 0;
    constraint define 3 [3]ind = 0; constraint define 4 [3]blknth = 0;
    constraint define 5 [3]scih = 0; constraint define 6 [3]bush = 0;
    constraint define 7 [3]subsci = 0; constraint define 8 indhe;
    constraint define 9 indme; constraint define 10 blkme;

```
/* Now mlogit equation will be estimated with constraints 1–10
   imposed, Table 3.3.
Note: constr(1–10) below imposes constraints */;
mlogit y age age2 blk blkovs ind indovs ovsbn north indnth
   blknth south indsth blksth highed indhe blkhe mided indme
   blkme subbus subsci bush scih, constr (1–10) base(1);
lrtest; /* Likelihood ratio test carried out: zero restrictions
   not rejected with chi2(13) = 8.1 */;
/* Predicted probabilities for EACH PERSON for outcomes
   1, 2, 3 are stored, respectively, in p1, p2, p3 */;
predict p1, outcome(1); predict p2, outcome(2); predict p3,
   outcome (3);
/* Predicted probabilities are summarized for: all men, whites,
   blacks, Indians (Table 3.4, upper panel) */;
summarize p1 p2 p3; summarize p1 p2 p3 if blk == 0 &
   ind == 0; summarize p1 p2 p3 if blk == 1;
   summarize p1 p2 p3 if ind == 1;
drop p1 p2 p3; /* Variables are released for subsequent use */;
/* The value of Z is calculated for EACH PERSON for outcomes
   1, 2, 3 and stored in z1, z2, z3 */;
predict z1, outcome(1) xb; predict z2, outcome(2) xb; predict
   z3, outcome(3) xb;
/* The means of z1, z2, z3 are stored in zm1, zm2, zm3 */;
summarize z1, mean; gen zm1 = r(mean); summarize z2, mean;
   gen zm2 = r(mean); summarize z3, mean; gen zm3 = r(mean);
/* The probabilities of outcomes 1, 2, 3 are calculated using
   Equations 3.12a and 3.12b, for all men */;
gen sum = 1 + exp(zm2) + exp(zm3); gen p1 = 1/sum; gen
p2 = exp(zm2)/sum; gen p3 = exp(zm3)/sum;
summarize p1 p2 p3; /* Predicted probabilities are shown for all
   men (Table 3.4, lower panel) */;
drop sum p1 p2 p3 zm1 zm2 zm3; /* Variables are released for
   subsequent use */;
/* Now computing the mean of z1 z2 z3 over white men and
   storing in zm1, zm2, zm3 */;
summarize z1 if blk == 0 & ind == 0, mean; gen zm1 = r
   (mean); summarize z2 if blk == 0 & ind == 0, mean;
gen zm2 = r(mean); summarize z3 if blk == 0 &
```

ind == 0, mean; gen zm3 = r(mean);
/* The probabilities of outcomes 1, 2, 3 are calculated using
Equations 3.12a and 3.12b, for white men */;
gen sum = 1 + exp(zm2) + exp(zm3); gen p1 = 1/sum; gen
p2 = exp(zm2)/sum; gen p3 = exp(zm3)/sum;
summarize p1 p2 p3 if blk == 0 & ind == 0; /* Predicted
probabilities are shown for white men (Table 3.4, lower
panel) */;
drop sum p1 p2 p3 zm1 zm2 zm3; /* Variables are released for
subsequent use */;
/* Now computing the mean of z1 z2 z3 over black men and
storing in zm1, zm2, zm3 */;
summarize z1 if blk == 1, mean; gen zm1 = r(mean);
summarize z2 if blk == 1, mean; gen zm2 = r(mean);
summarize z3 if blk == 1, mean; gen zm3 = r(mean);
/* The probabilities of outcomes 1, 2, 3 are calculated using
Equations 3.12a and 3.12b, for black men */;
gen sum = 1 + exp(zm2) + exp(zm3); gen p1 = 1/sum; gen
p2 = exp(zm2)/sum; gen p3 = exp(zm3)/sum;
summarize p1 p2 p3 if blk == 1; /* Predicted probabilities
are shown for black men (Table 3.4, lower panel) */;
drop sum p1 p2 p3 zm1 zm2 zm3; /* Variables are released for
subsequent use */;
/* Now computing the mean of z1 z2 z3 over Indian men and
storing in zm1, zm2, zm3 */;
summarize z1 if ind == 1, mean; gen zm1 = r(mean);
summarize z2 if ind == 1, mean; gen zm2 = r
(mean); summarize z3 if ind == 1, mean; gen zm3 = r(mean);
/* The probabilities of outcomes 1, 2, 3 are calculated using
Equations 3.12a and 3.12b, for Indian men */;
gen sum = 1 + exp(zm2) + exp(zm3); gen p1 = 1/sum; gen
p2 = exp(zm2)/sum; gen p3 = exp(zm3)/sum;
summarize p1 p2 p3 if ind == 1; /* Predicted probabilities
are shown for Indian men (Table 3.4, lower panel)*/;
drop sum p1 p2 p3 zm1 zm2 zm3; drop z1 z2 z3; /* Variables are
released for subsequent use */;
/* Ethnic simulations follow */;
gen blko = blk; gen indo = ind; /* Ethnic variables

```
saved */;
replace blk = 0; replace ind = 0; /* Everyone is white */;
/* All interaction variables involving blk and ind need to be
   recomputed */;
replace indnth = ind*north; replace indsth = ind*south;
replace blknth = blk*north; replace blksth = blk*south;
replace blkovs = blk*ovsbn; replace indovs = ind*ovsbn;
replace blkhe = blk*highed;
/* Predicted probabilities for EACH PERSON for outcomes
   1, 2, 3 are stored, respectively, in p1, p2, p3 */;
predict p1, outcome(1); predict p2, outcome(2); predict p3,
   outcome(3);
summarize p1 p2 p3; /* Predicted probabilities are summarized
   over entire sample:
EVERYONE assumed white (Table 3.5, upper panel) */;
drop p1 p2 p3;
/* The value of Z is calculated for EACH PERSON for
   outcomes 1, 2, 3 and stored in z1, z2, z3: EVERYONE
   assumed white */;
predict z1, outcome(1) xb; predict z2, outcome(2) xb; predict
   z3, outcome(3) xb;
/* The mean of z1, z2, z3 are computed over entire sample and
   stored in zm1, zm2 zm3 */;
summarize z1, mean; gen zm1 = r(mean); summarize z2,
   mean; gen zm2 = r(mean); summarize z3, mean; gen
   zm3 = r(mean);
/* The probabilities of outcomes 1, 2, 3 are calculated using
   Equations 3.12a and 3.12b:
EVERYONE assumed white */;
gen sum = 1 + exp(zm2) + exp(zm3); gen p1 = 1/sum; gen
   p2 = exp(zm2)/sum; gen p3 = exp(zm3)/sum;
summarize p1 p2 p3; /* Predicted probabilities are summarized
   over entire sample:
EVERYONE assumed white (Table 3.5, lower panel) */;
drop z1 z2 z3 sum p1 p2 p3 zm1 zm2 zm3; /* Variables are
   released for subsequent use */;
replace blk = 1; /*Everyone is black */;
/* All interaction variables involving blk and ind need to be
```

```
recomputed */;
replace blknth = blk*north; replace blksth =
   blk*south;
replace blkovs = blk*ovsbn; replace blkhe = blk*highed;
/* No need to replace ind* variables since ind is already 0 */;
/* predicted probabilities for EACH PERSON for outcomes
   1, 2, 3 are stored, respectively, in p1, p2, p3 */;
Predict p1, outcome(1); predict p2, outcome(2); predict p3,
   outcome(3);
summarize p1 p2 p3; drop p1 p2 p3; /* Predicted probabilities
   are summarized over entire sample:
EVERYONE assumed black (Table 3.5, upper panel) */;
/* The value of Z is calculated for EACH PERSON for outcomes
   1, 2, 3 and stored in z1, z2, z3: EVERYONE assumed black */;
predict z1, outcome(1) xb; predict z2, outcome(2) xb; predict
   z3, outcome(3) xb;
/* The mean of z1, z2, z3 are computed over entire sample and
   stored in zm1, zm2 zm3 */;
summarize z1, mean; gen zm1 = r(mean); summarize z2, mean;
   gen zm2 = r(mean); summarize z3, mean; gen zm3 = r(mean);
/* The probabilities of outcomes 1, 2, 3 are calculated using
   Equations 3.12a and 3.12b:
EVERYONE assumed black */;
gen sum = 1 + exp(zm2) + exp(zm3); gen p1 = 1/sum;
   gen p2 = exp(zm2)/sum; gen p3 = exp(zm3)/sum;
summarize p1 p2 p3; /* Predicted probabilities are summarized
   over entire sample:
EVERYONE assumed black (Table 3.5, lower panel) */;
drop z1 z2 z3 sum p1 p2 p3 zm1 zm2 zm3; /* Variables are
   released for subsequent use */;
replace ind = 1; replace blk = 0; /*Everyone is
   Indian */;
/* All interaction variables involving blk and ind need to
   be recomputed */;
replace indnth = ind*north; replace indsth = ind*south;
replace blknth = blk*north; replace blksth = blk*south;
replace blkovs = blk*ovsbn; replace indovs = ind*ovsbn;
replace blkhe = blk*highed;
```

```
/* Predicted probabilities for EACH PERSON for outcomes
   1, 2, 3 are stored, respectively, in p1, p2, p3 */;
predict p1, outcome(1); predict p2, outcome(2); predict p3,
   outcome(3);
summarize p1 p2 p3; drop p1 p2 p3; /* Predicted probabilities
   are summarized over entire sample:
EVERYONE assumed Indian (Table 3.5, upper panel) */;
/* The value of Z is calculated for EACH PERSON for
   outcomes 1, 2, 3 and stored in z1, z2, z3: EVERYONE
   assumed Indian */;
predict z1, outcome(1) xb; predict z2, outcome(2) xb;
   predict z3, outcome(3) xb;
/* The mean of z1, z2, z3 are computed over entire sample
   and stored in zm1, zm2 zm3 */;
summarize z1, mean; gen zm1 = r(mean); summarize z2, mean;
   gen zm2 = r(mean); summarize z3, mean;
   gen zm3 = r(mean);
/* The probabilities of outcomes 1, 2, 3 are calculated
   using Equations 3.12a and 3.12b:
EVERYONE assumed
   Indian */;
gen sum = 1 + exp(zm2) + exp(zm3); gen p1 = 1/sum;
   gen p2 = exp(zm2)/sum; gen p3 = exp(zm3)/sum;
summarize p1 p2 p3; /* Predicted probabilities are summarized
   over entire sample:
EVERYONE assumed Indian (Table 3.5, lower panel) */;
drop z1 z2 z3 sum p1 p2 p3 zm1 zm2 zm3; /* Variables are
   released for subsequent use */;
replace blk = blko; replace ind = indo; /* Original
   variables restored */;
/* End of program */;
```

NOTES

1. Available at http://www.indiana.edu/˜statmath/stat/all/cat/giant.html.

2. With just two outcomes, ordinary logit and probit methods can be used irrespective of whether the dependent variable is ordinal or nonordinal.

3. However, ordinal models may be based on distributions other than the logit or probit. For example, the log-log or negative log-log or complementary log-log distributions for skewed ordinal data (see Agresti, 1990). The general point is that ordered logit and probit models are a subset of a much larger set of options of dealing with ordinal dependent variables. Indeed, if the number of outcomes is large (e.g., greater than 20) then the methods of ordered logit and probit could become cumbersome and it might be preferable to use other methods (see: Jöreskog and Sörbom, 1988, pp. 44–45, and 1993, pp. 1–17, for discussion of this point).

4. The ranking may be on the basis of income, in which case there is an objective hierarchy of occupations, or it may be based upon cultural assumptions and prejudice.

5. Better, that is, than treating outcomes as ordered, unless one had good reasons for not imposing a ranking.

6. Remembering that $N = N_1 + N_2 + N_3$.

7. Because of the manner in which they are computed, these estimates are termed, *maximum likelihood estimates.*

8. There are other approaches to modelling ordinal outcomes, such as adjacent categories logit, stereotype, logit and continuation ratio logit models (see Agresti, 1996, pp. 216–220).

9. It closely resembles a t distribution with seven degrees of freedom (Greene, 2000, p. 815).

10. See Brant (1990) for a discussion of this property.

11. Greene (2000) actually indexes his cutoff points beginning with 0. So in his notation, $\mu_0 = 0$. For ease of comparison, however, I have begun the index for his cutoff points at 1.

12. Remember that Greene (2000, p. 876) writes $\widehat{W}_i$ as $\beta'x$.

13. Since $\mu_1 = 0$, $\delta_1 = -\beta_0$, so that the first STATA cutoff point is the negative of the intercept.

14. Normally distributed with mean 0 and variance 1.

15. Remember that the three probabilities must sum to unity.

16. The coefficients associated with one of the outcomes have to be normalized for identifiability. This is discussed in the next chapter.

17. These individuals represent a 2% sample of the census records. For a fuller account of the data used see Borooah (2000a).

18. A person was regarded as not deprived if $D_i = 0$; as mildly deprived if $0 < D_i \leq \bar{D}$; as severely deprived if $D_i > \bar{D}$ where $\bar{D}$ was the mean of the deprivation index.

19. After normalization, $AGE_i = 1$, for those who were 17 years old; $AGE_i = 2$, for those who were 18 years old; and so on.

20. These are qualifications generally obtained at 18+.

21. These were: Belfast (Area1); Ards, Castlereagh, North Down (Area 2); Down, Lisburn (Area 3); Carrickfergus, Larne, Newtownabbey (Area 4); Antrim, Ballymena and Ballymoney (Area 5); Armagh, Newry & Mourne (Area 7); Coleraine, Cookstown, Maghrafelt, Moyle (Area 6); Banbridge, Craigavon, Dungannon (Area 8); Derry, Limavady (Area 9); and Fermanagh, Omagh, Strabane (Area 10). Because of

multicollinearity, all 10 areas cannot be included in Equation 2.16. $AREA_1$ (Belfast) is the area that was dropped (aliased) from the equation.

22. All non-Roman Catholics are identified as "Protestants," although this latter group contained persons who either did not state a religion (7.2% of Northern Ireland residents) or declared that they had no religion (3.8% of Northern Ireland residents).

23. Note that (a) under STATA, $\beta_1 = 0$, and (b) from Equation 2.16, $dD_i/dAGE_i = \beta_4 + 2\beta_5 AGE_i$

24. These ratios represent a form of the Wald statistic.

25. In order to test the parallel slopes assumption the model was also estimated using multinomial logit and this yielded a likelihood value (denoted L_2) of -12380.36. The computed value of $2(L_2 - L_1)$ was 86.4 (the comparison was with the ordered logit model) and *on a strict likelihood-ratio test interpretation* this exceeded the 5% critical $\chi^2(20)$ value of 31.41. However, as noted earlier, the value of $2(L_2 - L_1)$ is only suggestive since the statistic does not provide the basis for a likelihood-ratio test.

26. Note: $xb + u$ (in the Tables) $= \widehat{Z}_i + \varepsilon_i$ (in the text).

27. The cutoff points are really only coefficients of the model and are estimated along with the slope coefficients using maximum likelihood methods.

28. These proportions were set out at the start of this section.

29. Though the outcome would be different if the average was defined as the median (rather than the mean) of the individual probabilities.

30. Note that under STATA $\beta_1 = 0$.

31. For ease of presentation, the subsequent discussion is entirely in terms of the logit model.

32. For example, nearly 16% of Catholics, compared to 8% of Protestants, in the sample were single parents.

33. Note that these predicted means would be identical to the sample proportions of Catholics and Protestants at the different deprivation levels.

34. That is, Protestant attributes evaluated at Protestant coefficients.

35. In the case of unemployed and retired persons these probabilities hardly changed.

36. In the sense of having the same average probability of being mildly deprived.

37. These individuals represent a 2% sample of the census records. For a fuller account of the data used see Borooah (2000b).

38. The default was no post-18 qualifications.

39. Relevant only if the person had post-18 qualifications. The default area was Arts-related subjects.

40. The default group was white.

41. The default area was London.

42. In terms of the standard regions of Britain: East Midlands; East Anglia; the Southeast (excluding London); the Southwest.

43. The North; Yorkshire; West Midlands; Northwest; Wales; Scotland.

44. Note that any one of the three outcomes could have been chosen as the base outcome.

45. Of course, if the relevant coefficient for a particular group was positive, then it would enjoy an "ethnic bonus."

46. These ratios represent a form of the Wald statistic.

47. Other than age: older persons were more likely to be in the higher occupational classes or, in terms of the estimated coefficients of Equation 3.14, $\hat{\theta}_{j1} > 0$, $j = 1, 2$.

48. Unless there is a specification to the contrary, the discussion in this section and subsequent sections always carry a ceteris paribus clause.

49. Though not for blacks in the North in PMT employment.

50. In other words, 41.5% of all men, 41.8% of white men, 29.3% of black men, and 35.1% of Indian men in the sample were in PMT jobs.

51. By replacing $\overline{Z}_j$ in the equations with, as appropriate, $\overline{Z}_{Wj}$, $\overline{Z}_{Bj}$ and $\overline{Z}_{Ij}$ ($j = 1, 2, 3$).

52. Or equivalently calculated using the mean values $\overline{X}_{Wr}$ of the individual values of the determining variables X_{ik} for the white persons in the sample.

53. It should be noted that this section presents simulation-based strategies concerned with a specific set of questions relating to differences in intergroup outcomes. For a more general approach to simulation strategies see Tomz, Wittenberg, and King (1999). Tomz et al. (1999) have developed a program, Clarify, that uses stochastic simulation to convert the raw output of statistical procedures into results that are of direct interest to researchers. This program is designed for use with STATA.

54. A similar exercise can be performed for the hypothetical case where everyone in the sample is Indian.

55. Which were α_{jr} for whites and $\alpha_{jr} + \gamma_{jr}$ for Indians.

56. The $\widehat{Z}_{ij}$ are computed using the estimates shown in Table 3.3 in conjunction with the values of the determining variables for every individual. When everyone is white, $IND_i = BLK_i = 0$ and when everone is black, $BLK_i = 1$, in the estimated Equation 3.14.

57. On the other hand, if $p_j^B > p_j^W$, then $\lambda_j^B > 1$ and there is an ethnic advantage.

58. On the other hand, if $s_j^B > s_j^W$, then $\mu_j^B > 1$ and there is an overall bonus.

59. This depends only on the ratio of sample proportions which, of course, remain unchanged.

60. In the multinomial model, log-odds-ratio depended solely on the (differences between) coefficients associated with j and k and was independent on the coefficients associated with any other outcome (Equation 3.6).

61. Domencich and McFadden (1996).

62. This drawback is shared by many empirically convenient functional forms in economics, for example the Cobb-Douglas or the constant elasticity of substitution (CES) functional forms.

63. Though it must be emphasized that new alternatives could be introduced by splitting existing ones: for example, skilled manual/nonmanual into skilled manual and skilled nonmanual.

64. See the discussion earlier in the section on how probabilities are recomputed when a new alternative is introduced.

65. Available at http://www.indiana.edu/~statmath/stat/all/cat/giant.html.

References

AGRESTI, A. (1996). *An introduction to categorical data analysis.* New York: John Wiley.

ALDRICH, J. H., and NELSON, F. D. (1984). Linear probability, logit, and probit models, *Quantitative Applications in the Social Sciences, 07–045.* Beverly Hills, CA: Sage.

BEN-AKIVA, M., and LERMAN, S. (1985). *Discrete choice analysis.* London: MIT Press.

BOROOAH, V. K., and CARCACH, C. (1997). Fear and crime: Evidence from Australia. *British Journal of Criminology, 37,* 634–656.

BOROOAH, V. K. (2000). Targeting social need: Why are deprivation levels in Northern Ireland higher for Catholics than for Protestants? *Journal of Social Policy, 29,* 281–301.

BOROOAH, V. K. (2001). How do employees of ethnic origin fare on the occupational ladder in Britain? *Scottish Journal of Political Economy, 48,* 1–26.

BRANT, R. (1990). Assessing proportionality in the proportional odds model for ordinal logistic regression. *Biometrika, 44,* 131–140.

CRAMER, J. (1999). Predictive performance of the binary logit model in unbalanced samples. *Journal of the Royal Statistical Society, Series D (The Statistician), 88,* 85–94.

DEMARIS, A. (1992). Logit modelling. *Quantitative Applications in the Social Sciences, 07–086.* Newbury Park, CA: Sage.

DESAI, M., and SHAH, A. (1988). An econometric approach to the measurement of poverty. *Oxford Economic Papers, 40,* 505–522.

DOMENCICH, T. A., and McFADDEN, D. (1996). *Urban travel demand: A behavioral analysis.* Amsterdam: North-Holland.

GREENE, W. H. (1995). *LIMDEP, version 7.0: User's manual.* Bellport, NY: Econometric Software.

GREENE, W. H. (2000). *Econometric analysis.* Englewood Cliffs, NJ: Prentice Hall (4th ed.).

HAUSMAN, J., and McFADDEN, D. (1984). A specification test for the multinomial logit model. *Econometrica, 52,* 1219–1240.

HENSHER, D. (1986). *Simultaneous estimation of hierarchical logit mode choice models* (Working Paper No. 24). MacQuarie University, School of Economic and Financial Studies.

JÖRESKOG, K., and SÖRBOM, D. (1988). *PRELIS: A program for multivariate data screening and data summarization.* Chicago: Scientific Software Inc.

JÖRESKOG, K., and SÖRBOM, D. (1993). *LISREL 8: Structural equation modeling with the SIMPLIS command language.* Hillsdale, NJ: Lawrence Erlbaum.

KAY, R., and LITTLE, S. (1986). Assessing the fit of the logistic model: A case study of children with haemolytic uraemic syndrome. *Applied Statistics, 35,* 16–30.

LIAO, T. F. (1994). Interpreting probability models. *Quantitative Applications in the Social Sciences, 07–101.* Newbury Park, CA: Sage.

McFADDEN, D. (1973). Conditional logit analysis of qualitative choice behavior. In P. Zarembka (Ed.), *Frontiers in econometrics.* New York: Academic Press.

MENARD, S. (1995). Applied regression analysis. *Quantitative Applications in the Social Sciences.* Thousand Oaks, CA: Sage.

NOLAN, B., and WHELAN, C. T. (1996). *Resources deprivation and poverty.* Oxford: Clarendon Press.

96

PIACHAUD, D. (1987). Problems in the definition of poverty. *Journal of Social Policy,* *16,* 147–164.

SCHMIDT, P., and STRAUSS, R. P. (1975). The Prediction of occupation using multiple logit models. *International Economic Review, 13,* 471–486.

STATA (1999). *Stata reference manual release 6.* College Station, TX: Stata Press.

TOMZ, M., WITTENBERG, J., and KING, G. (1999). *CLARIFY: Software for interpreting and presenting statistical results, version 1.2.1.* Cambridge, MA: Harvard University Press. (http://gking.harvard.edu/).

TOWNSEND, P. (1979). *Poverty in the United Kingdom.* Harmondsworth: Penguin.

VEALL, M. R., and ZIMMERMANN, K. F. (1996). Pseudo-R^2 measures for some common limited dependent variable models. *Journal of Economic Surveys, 10,* 241–260.

ABOUT THE AUTHOR

VANI K. BOROOAH is Professor of Applied Economics at the University of Ulster, a position he has held since 1987. Prior to that he was Senior Research Officer at the Department of Applied Economics at the University of Cambridge and Fellow and College Lecturer in Economics at Queens' College, Cambridge. Born and mostly educated in India, Borooah earned his PhD from the University of Southampton and his MA from the University of Bombay. His research focuses on poverty, inequality, and labor market outcomes, and he has published several academic papers and books on these subjects. He is particularly interested in the public policy implications of intergroup differences in economic and social outcomes. Borooah combines his study of the policy implications of intergroup differences with an interest in the economies and societies of developing countries. He has been President of the European Public Choice Society and President of the Irish Economics Association.

Printed in the United States
By Bookmasters